Federico Moreno

**Petroleum in Peru**

From an Industrial Point of View

Federico Moreno

**Petroleum in Peru**
*From an Industrial Point of View*

ISBN/EAN: 9783743358072

Manufactured in Europe, USA, Canada, Australia, Japa

Cover: Foto ©ninafisch / pixelio.de

Manufactured and distributed by brebook publishing software (www.brebook.com)

Federico Moreno

**Petroleum in Peru**

# PETROLEUM IN PERU,

## FROM AN INDUSTRIAL POINT OF VIEW.

BY

## FEDERICO MORENO,

EX-PREFECT OF PIURA.

Dedicated to the "Sociedad Geográfica de Lima"

[Translated from the original Spanish.]

LIMA

F. Masias & Co. Printers.
317, Union Street.

1891.

# PETROLEUM IN PERU,

## FROM AN INDUSTRIAL POINT OF VIEW.

BY

## FEDERICO MORENO,

EX-PREFECT OF PIURA.

Dedicated to the "Sociedad Geográfica de Lima"

[Translated from the original Spanish.]

LIMA

F. Masias & Co. Printers.
317, Union Street.

1891.

# INDEX.

|  | PAGES. |
|---|---|
| Dedication | 3 |

## PART I.

| | |
|---|---|
| History | 5 |
| Extension of the beds | 10 |
| Scientific opinions in favor of Peruvian petroleum | 15 |
| Markets for consumption | 24 |
| Its future as fuel | 34 |
| Works already established | 39 |
| Fields which are being explored | 46 |
| General aspect of the district | 55 |

## PART II.

| | |
|---|---|
| Industrial development of petroleum in the United States of North America | 63 |
| Flowing and Spouting Wells | 69 |
| Prospecting for oil | 73 |

PAGES.

Russian petroleum fields........................... 77
Petroleum deposits, Refineries, Exportation............................................... 84
Transportation of kerosene....................... 86
Petroleum worship................................. 89
Monopoly ............................................. 92

## PART III.

Scientific aspect, origin and chemical composition of petroleum.................. 95
Where petroleum is found........................ 99
Refinement ............................................ 104

## PART IV.

Combustible oil....................................... 113
Conference in London............................. 117
Report of E. P. Larkin, Esq.,................... 139
   ,,   ,,  W. Warren, Esq...................... 157

*Lima, August 1st. 1891.*

Mr. President:

As the petroleum districts of Peru, according to the reports of eminent engineers, have been shown to be the second in importance in the world, I have come to the conclusion that it is a matter of national interest to publish this little work, the object of which is simply to show the country the new, and perhaps inexhaustible, source of wealth it possesses, and which, if its industrial development is undertaken with prudence and intelligence, may prove highly profitable.

I beg of you, Mr. President, to accept, with your habitual benevolence, this deficient study, which I have the honor to dedicate to the "Sociedad Geográfica de Lima," as a mark of respect.

With sentiments of the highest consideration, I subscribe myself, Mr. President.

Your most obedient servant.

F. Moreno.

To Dr. Luis Carranza, President of the "Sociedad Geográfica de Lima."

# PETROLEUM IN PERU,

### FROM AN INDUSTRIAL POINT OF VIEW.

## PART FIRST.

### CHAPTER I.

*Its History.*

Although the narrow columns of a newspaper do not suffice for a long disertation on the importance of this substance, which to day attracts the serious attention of European governments and capitalists, we believe an opportune moment has arrived to make a brief study of this new source of wealth, now that Peru turns out to be one of the few privileged nations, as a petroleum producing country.

For more than a century past, people here have known *Brea* or *Copé*, [asphaltum] an almost exclusive product of the *hacienda* of Pariñas, in the district of Amotape, in the Department of Piura, and which was employed then, as it is now, to coat the insides of *botijas, piscos* [earthenware receptacles of about 150 lb. and 25 lb. capacity, respectively] and other articles of the kind, for the ardent spirits made in Ica, Locumba and Moquegua, in order to give them a special flavor, preserve them better, and prevent injurious fermentations.

In the memorials of the Vice-roys of Peru, an entry is found, in favor of the Spanish Crown, of $ 28,000 for the export tax on this substance.

Somewhere in the neighbourhood of 1860, the happy idea occurred to Manuel Antonio de Lama to refine petroleum, for the first time in Peru, taken from the asphaltum beds of Máncora and Zorritos, the property, at that time, of the wealthy Don Diego de Lama.

Subsequent to that date, a large commercial house, of Lima, established expensive works at Negritos, from the wells of which place an abundance of oil rose to a height of seventy feet above the ground, the whole of which was lost for want of receptacles for its deposit. This *contre temps*, as well as others met with by the house of Rollin Thorne, the first to establish works of any extent in the oil fields of the province of Paita, was the cause of their abandoning their idea, at a heavy loss, due exclusively to the fact that the petroleum industry and its various applications were still in their infancy, only a few years having passed since the discovery of this oil in the United States, where kerosene for illuminating purposes was first made.

Don Alejandro Rudens, a merchant of Paita, and Consul of the United States, at that time, associated with other persons, organised a big company, with a capital of $ 4.000,000, for the working, on a large scale, of the important oil fields that had been discovered; but, owing to the civil war in the United States, the scheme fell through.

In 1876 there arrived in Peru the wealthy Mr.

Prentice, a proprietor of extensive oil works in Pennsylvania, who, on convincing himself of the richness of the supply in Tumbez, after sinking a well 500 feet deep, from which a heavy stream of oil spouted forth, asked the Government for the concession to him of the entire petroleum district; but this request was not granted, it being contrary to the mining ordinances and other laws of the country. Mr. Prentice, after the Government's negative, left the country and abandoned his project.

This is the true history of petroleum in Peru.

From the studies made in that epoch, in the district of Zorritos, it will be seen: that of the two wells which were opened, one gave 60 barrels per day; and the other, at a depth of 500 feet, gave as much as 600 barrels per day; and it was the unanimous opinion of the engineers that this last well, if sunk a hundred metres more, would yield 5,000 barrels per day.

Eminent English and American engineers visited that region then and later on, and arrived at the conclusion, from the studies made, that the petroleum fields on the coast of the Department of Piura were the second in importance in the world.

As a matter of fact, the petroleum deposits, commonly called *aceite de piedra* [stone oil] or *kerosene*, situated along the coast of the vast Department of Piura, constitute, on account of their extent, excellent quality and proximity to the sea (100 to 16,000 metres) an immense source of wealth, the whole of which will be available once the deposits of oil in the United States, in the natural order of things, are exhausted, and when petroleum, solving a great economic problem, as it is, in fact, already solved, is definitely employed as the best combustible known, in all steam-engines, and in steamers as well as in locomotives, as is done at present, in Russia, with an economy over coal of 50%; for there is no application, in the form of combustible, which cannot be given to petroleum; while the cost of its extraction is as low as the expense of taking water from a well.

Its industrial applications, at present, are many, and they will increase, in future, in proportion to the progress of the studies constantly being made, according to articles published daily by the European press, to discover further applications of this valuable substance.

Nearly all the locomotives on our railways use petroleum, and it will not be long before we see the same thing in the merchant steamers plying along the coast.

The Peruvian Corporation, in the last report presented to the shareholders, says, regarding this combustible: "Experiments made on the Oroya Railway and also on a few others, with the object of testing the efficacy of petroleum as a combustible instead of coal, have so far given such satisfactory results that arrangements have been made with the London and Pacific Petroleum Company, established in the North of Peru, for them to furnish the oil necessary for this and other roads. The economy in the cost of combustible, to judge from the results of the experiments, will amount to over 40 per cent."

Similar trials, with magnificent results, have lately been made in some of the nitrate *oficinas* in Tarapacá, for which industry the application of petroleum as a combustible would be its salvation, English coke, placed on the pampas of Iquique, being so dear.

The use of petroleum in the machinery on the

sugar estates in the North may be considered as of the utmost importance, leaving the *bagazo* (pressed sugar-cane) which is to day used as a combustible, to be applied to another industry.

Captain N. V. Coxmichael, in the interesting report read at a conference held in London on February 13$^{th}$ of this year, declares, à *propos* of the combustible oil, according to his own experience, that merchant steamers which consume petroleum instead of coal will, apart from other advantages which he enumerates, obtain an economy of £ 7,000 per annum.

The Pacific Steam Navigation Company, therefore, which owns twenty vessels, would economise the sum of £ 140,000 per annum, or 980,000 sols of our money!

According to the opinion expressed by Mr. Nelson Boyd, the most conspicuous of petroleum mining engineers, and who at present resides in London, there is petroleum enough in the beds discovered so far, for many centuries; whereas the coal, according to measurements taken in the coal mines of England and Belgium, will hardly suffice for sixty years, while it is getting dearer every day.

The numerous factories in the neighborhood of Chicago alone, by using petroleum, save annually 500,000 tons of coal, which they formerly consumed.

We do not know the exact cost of the extraction of the crude oil; but, in view of the fact that, in order to convert it into a combustible, by removing its inflammable qualities, it is only necessary to expose it in the open air for a month, or submit it, immediately after its extraction from the wells, to a temperature of 350°, we have every reason to believe that this operation will not be too costly, although this would not be the case with the residue, which is only obtained in the proportion of 10 to 30 per cent, after the refinement of the oils and the extraction of the kerosene.

From the foregoing it will be seen: that petroleum, from an economical standpoint, is to coal, as beet-root sugar to sugar made from sugar cane. In its employment on board men-of-war, it differs from coal, as the modern repeating rifle, with smokeless powder, differs from the old fire-locks. And its power as a generator of heat, is to coal what dynamite is to powder. In a word, petroleum is the sepulchre of coal.

To recapitulate:

The industrial applications which have been given petroleum, for the moment, are:

$1^{st.}$ A generator of steam in merchant steamers and men-of-war.

$2^{nd.}$ The same thing in railway locomotives, and all kinds of steam engines.

$3^{rd.}$ As fuel for smelting ores, especially silver, lead and copper.

$4^{th.}$ As fuel for the working of nitrate or any other substance.

$5^{th.}$ For the heating of ovens for baking bread.

$6^{th.}$ For the heating of rooms in cold countries, using the hardened oil, as is done in Russia.

$7^{th.}$ For the fabrication of illuminating gas in cities.

$8^{th.}$ For illuminating purposes in the form of kerosene.

## CHAPTER II.

*Extension of the Beds.*

According to the opinion expressed by the eminent North American engineer E. P. Larkin, in a pamphlet published in New York in 1866, the petroleum beds visited by him in the Department of Piura, comprise an area of 800 square leagues, or 7,200 square miles, the equivalent of 4,408,000 English acres; but, according to Mr. Prentice, the length is 120 miles, and the width 60, giving a total area of 1,000,000 hectaras, the equivalent of 250,000 claims of 40,000 square metres each, according to our mining laws.

The district to which these engineers refer is merely that comprised between Cabo Blanco, fifty miles north of Paita, and the extreme north of the province of Tumbez; but the late discoveries of the engineers Blume and Dávila, in the neighborhood of Punta de Aguja, in the district of Sechura, comprising Punta Falsa, Garita, Reventazon, Colorado, Nonura, Pisura, Bichayo, Salinas and other points such as Brea, Talara, Negritos, Playa Grande, Playa Chica, Pericos and Punta Foca, of the province of Paita, form an area equal in extent to the other.

The petroleum beds of Paita, therefore, comprise to-day, beyond any doubt, an area of 16,000 square miles, inferior, certainly, in extent, to the Russian petroleum beds, which comprise some 200,000 square miles, but very superior to those in the United States, which are hardly 120 miles long, with a variable width of from one to twenty. These furnish the commercial world, in the form of illuminating oil, or kerosene, 40,000,000 barrels per year, the annual output from 1860 to 1884 having been 266,000,000 barrels, the value of which is estimated at 443,991,380 dollars.

The area of the petroleum fields indicated by the engineers Larkin and Prentice may possibly, in reality, be greater, considering the fact that the Department in which they are situated comprises an extension of 70,463 square kilometres, the greater part of the same being on the coast, and of the same geological formation.

This formation of the three petroleum regions of the Centre, South and North, comprised in the Department of Piura, is identical, and only varies, at some places, in the outward appearances of the oil, either in its liquid state, or hardened by atmospheric action.

Thus we see that in Negritos, the region of the Centre, Sechura, the region ef the South, and Tumbez and Máncora, the region of the North, petroleum, at a depth of from 100 to 350 feet (English), is found with slight variations in quality. In Sechura, it is found, as a rare exception, in combination with gas, from a depth of one metre onwards, the temperature augmenting until it becomes insupportable.

As a general rule, in the petroleum fields, the earth on the superfice is of modern deposit. The oil is commonly found underneath, sometimes conglomerated with stone, and at others covered with a layer of sandy clay, mixed with gravel. The geological strata are the same, and are found in the following order:

$1^{st.}$ At the surface, a layer of sand varying from 0.25 to 4 metres in thickness.

$2^{nd.}$ White sandstone, hard in some places, and porous in others, the thickness varying from 0.30 to 2 metres. Where the sandstone is only slightly porous, the petroleum filters through and stains in; but where it is less compact, the filtration is abundant.

$3^{rd.}$ A layer of wet sand, soft and very fine, from 8 to 10 metres thick.

$4^{th.}$ Conglomerate of decomposed carbonate of lime, formed by the conglomeration of sea shells.

5th. Slate impregnated with oil, through the fissures of which the oil and gas escape.

From the formation of the strata, in will be seen that the work of drilling the wells for the extraction of the oil is very easy. As Mr. Warren, the engineer, said, in his report on Talara: "The operation of sinking wells is very easy on account of the softness of the ground and the perfection of modern apparatus. A well 200 feet deep was drilled in five days, and another, 345 feet in depth in ten days. The average depth in North America and Russia is 1,000 feet, and the cost per foot is much more.

"Everything in this place tends to prove the existence of petroleum, on a large scale, which can be obtained at different points. At one of these, Negritos, for example, twelve wells were sunk within a radius of two square miles, or 1280 acres, and petroleum was found in every one of them, in varying proportions. The wells which have been worked yield the some quantity of oil as those which have not been worked, and are situated about twelve miles to the East of a point called Brea, which is equally productive. Among the latter and those of Negritos, there is a petroleum field still unexplored.

"The area of the petroleum field of Negritos being two square miles, and calculating only the half of it to be productive, it would equal an area of 1229 acres, in Russia, which produced 14.375,000 barrels in 1886. The half of this alone, representing over 7.000,000 barrels, would be worth £ 1.100,000.

"If the quantity of oil yielded by it were only half that produced in the works on an area of the same size at Venango, in the United States, which yielded, during the same year 1,800,000 barrels, valued at £ 300,000, that of itself would be a satisfactory result for the Company. The different wells opened ta Brea, however, show, on an average, a result far superior to that obtained in the United States, or even in Russia, their output being *ten times* greater than that of the American wells.

"The possible yield of the two square miles of the petroleum field in Negritos may be estimated at from fifteen to eighteen million barrels, worth £ 3.000,000.

"The petroleum fields in the United States represent an area of 1339 square miles. Within a space there of only 37½ miles, 22,524 wells have been opened, and from these the greater part of the yield has been obtained.

"In Rusia, it is more or less the same. Within a space of 1229 acres, or in an area less than two miles square of the total petroleum field, which, at Baku, is estimated at about 1200 miles, there are 400 wells, which produced, and still produce, the greater part of Russian petroleum.

"The average product per square mile in Pennsylvania, worked up to the year 1885, may be estimated at 740,000 barrels; while in Russia the same area yields an immeasureably greater quantity.

"Judging from the results already obtained, the yield, per square mile, of the wells at Brea will be much greater than that obtained, up till to-day, in the United States, for an equal area.

"The average yield of petrolenm per well, in the United States, is calculated to be 12,000 barrels. In Russia, where it is much greater, it is estimated at 76,000.

"Estimating the yield of two square miles in the petroleum district between Negritos and the sea shore at 50,000 barrels per well, which is no exaggeration, we would have, with 300 wells, an output of 15,000,000 barrels, worth £3,000,000 for every two miles. Around these two miles, there is an area ten times greater in extent, and very rich in petroleum.

Fifty wells may be sunk in this place within a year, and would yield a good output of crude or refined petroleum.

"The crude oil would prove more profitable than the refined petroleum or kerosene, because it could be employed as fuel on the greater part of the North and South American coast, where English coal is so dear.

"The quality of the petroleum is, generally speaking, good, and the product of the last well sunk gives 91 $\frac{1}{2}$ per cent of oil, the greater part of which may be clarified like kerosene."

The foregoing report proves the great facility for sinking wells in those places, and the great abundance of petroleum.

Let us now see what is said about the quality of Peruvian petroleum by the celebrated chemists of the United States.

## CHAPTER III.

### *Scientific Opinions in Favor of Peruvian Petroleum.*

Dr. Salathé, in a report dated at Titusville, Pennsylvania, March 31st, 1885, referring to Peruvian petroleum, says: "The crude oil of Peru differs essentially from the petroleum of Pennsylvania. It may be considered as a product decomposed by heat, thus forming a series of hydro-carburets amongst the greasy and aromatic series. The odour is similar to that of the products of coal-tar.

"The density of oil this crude is 0.8480 at 15° C.,

or 36° B., by the hydrometer of the Standard Petroleum Association.

"Its analysis gives the following results:

Carbon..................84. 9 per cent.
Hydrogen..............13. 7 „
Oxygen................. 1. 4 „

"The heat of the flame is equivalent to 13,672 calories; the co-efficient of dilatation being 0.00072.

"The distillation of the crude oil has given the following results:

| Degrees Celsius. | Nos. Products. |
|---|---|
| 10 to 100, | 2.8—1 Cicachena, Pigolena. |
| 30 „ 80, | 9.0—2 Gasoline |
| 80 „ 150, | 11.1—3 Benzine |
| 150 „ 230, | 18.5—4 Kerosene light. |
| 230 „ 280, | 10.0—5 Kerosene, heavy. |
| At high temperatures | 12.8—6 Lubricating oil, light. |
|  | 4.8—7 Lubricating oil, heavy |
|  | 31.0—8 Asphaltum. |

"All the light products that are distilled at a temperature of between 10 to 15 degrees Celsius, are distinguished by a very agreeable odour, while the light oils of the petroleum in Pennsylvania have a *detestable odour*, which makes it necessary for them to be *disinfected.* So far as the oils which are distilled at a temperature of between 150° to 280° are concerned, a *small quantity* of sulphuric acid is sufficient to purify them, a barrel containing 42 gallons of oil requiring only 2½ to 3 lbs. of acid.

"The precipitate that is formed, when a sufficient quantity of acid has been added, *is not black as it is* in *our Pennsylvania oils.* The acid gives a *reddish brown* colour, *proving* that there is not the same quantity of tar in Peruvian oils that there is in *ours*

"The lubricating oils that I have separated and marked Nos. 6 and 7 are distinguished by the absence of *parafine*, which is *never missing* in the heavy Pennsylvania oils, and which makes it *very difficult* to obtain *good* lubricating oils from them.

"Samples Nos. 6 and 7 have been exposed to a very low temperature (33 degrees below zero Celsius) without becoming solidified, merely acquiring the consistency of syrup. Such properties give *great value* to these heavy petroleum oils, assimilating them to the lubricating oils of Russia, which are considered the best for their purpose.

"The distillation may be terminated the moment that vapours of a yellowish orange colour begin to issue from the retort, and which it is difficult to condense. The residue forms a product something like asphaltum (No. 8) and may be used as such for street pavements.

"In case it is not desired to obtain asphalt, the distillation may be continued in separate retorts until coke (n°. 9) is formed, which is an excellent combustible, leaving no ashes and giving a high temperature".

So much for the report of the famous Dr Salathé.

Let us now hear the opinion of another chemist no less eminent:

*Boston, July 28th, 1886.*

To EDWARD Towks, Esq.,
            Payta, Perú, S. A.

Dear Sir:

Having carefully examined and made fractional distillations of the oil obtained near the sea, at Negritos, Perú, S. A., in March, 1885, I submit for your consideration the following report.

In March 1885, I visited the petroleum fields between the rivers Honda and Chira, in Peru, on the west coast of South America; and from one of the abandoned wells in Negritos I took ten gallons of oil, filled two iron drums of fire gallons each with it, and, after duly closing and sealing them, I shipped them to Boston, Mass. N. A., in order to examine the oil carefully.

On the arrival of the drums containing the oil, I sent it to the laboratory of the Downer Kerosene Company, South Boston, where I broke the seals in the presence of my brother, Mr. Joshua Merrill, the said Company's chemist. The seals had not been damaged on the voyage: they were intact. The oil was found to be of a dark olive green colour, and of an agreeable odor, comparatively free from water and dirt. Its weight reached 290 in the hydrometer of Beaumé, or 883 S. G. at 60° Fahrenheit, although the liquid was thin. It should be observed that the gravity of liquids does not always imply thickness of "body"; to which I must add that the samples in question were taken from a well in which they had remained exposed to atmospheric distillation, assisted by an almost constant current of air on account of the prevailing winds, and the oil must have gradually thickened. I do not doubt, therefore, that the oils of this place will be found a few degrees lighter when taken from recently opened wells.

I placed the oil in a still, and, applying sufficient, heat, obtained: 1st. Five degrees of naphtha, which appeared colorless in the condenser, and of a very mild odor, requiring no further treatment to prepare it for commerce; it being worthy of note that this crude naphtha is clearer than the best treated naphtha of the petroleum of Pennsylvania. 2nd. I obtained 50 per cent. of kerosene, or illuminating oil, as clear as water, 110° F. proof, and of a peculiarly sweet odor. This kerosene burned in ordinary lamps, giving a white and luminous flame, equal to that given by oils of the same quality in Pennsylvania, and might be reputed as first class. 3rd. After the kerosene, I obtained 10 per cent. of medium fine oil, 300° proof, of a lemon color, and a slight, agreeable odor. This oil is used, throughout this country, in factories, on railways and on board steamers, and is burned in lamps made specially for the purpose. The product obtained from the Negritos oil burns very well in such lamps.

I next obtained lubricating and fixed oil, all in a single product. It gave 30 per cent of a very remarkable, oil, slightly amber coloured, weighing 20° Beaumé, almost without odor and with a good "body", but liquid. It contained no parafine wax, and remained liquid at a temperature of ten degrees below zero, without congealing, and absorbing only a small quantity of oxygen.

*This oil is the best brilliant petroleum that I have ever seen*, and is of great value to lubricate all kinds of machinery, and soften leather, wool, etc. An oil like this, put on the market at a fair price, would be easily and immediately placed. I classify the *Negritos* oil among aromatic petroleums.

*Rufus S. Merrill.*

104 Water Street,
Boston, Mass.

The following document, and what was said by Mr. N. Coxmichael in the public conference in London, on February 13th last, to which we have already alluded, sufficiently proves the advantage and economy of the use of petroleum instead of coal.

OPINION OF A. D. BRYCE DOUGLAS ON THE SAME SUBJECT.

14 Great George Street—London, S. W.

*Barrow-in-Furness, May 18th, 1888.*

Dear Sir:

In reply to your request regarding my experience in connection with petroleum as fuel on board steamers in the Pacific, I have the pleasure to inform you that when I was Superintendent of the Factory of the Pacific Steam Navigation Company, I arranged one of their small steamers, the "Supe", to burn oil, which it did with very great success.

At that time (1874) the price of Cardiff coal, in Callao, was $.17 per ton, and we paid only 7 cents per gallon for oil.

From a memorandum of those trials, it appears that on a short voyage from Callao to Guañape, the quantity of oil consumed was 2200 gallons, which, compared with the 15 tons of coal that was burned before, was a great advantage in favor of the oil. Thus:

15 tons of coal at $.17............... $. 255
2220 gallons of oil at 7 cents........ 154

Difference in favor of oil.............. $. 101

The oil used then was from the wells near Paita

(Negritos) and for a long time past, I have been under the impression that an immense and, probably, inexhaustible supply of petroleum exists there. I think it even exists underneath the sea, because, on many occasions, I have seen the surface of the water covered with oil, for a number of miles around, in front of Point Pariñas.

I enclose you a copy of the original memorandum which I addressed to Mr. Petrie, General Manager of the Company on the Coast.

Your obedient servant,

*A. D. Brycc-Douglas.*

*To Herbert W. C. Tweddle, Esq.*

Notwithstanding the respected opinions which we have just copied in favor of oil as fuel, we will transcribe some paragraphs from a notable work published in Paris, in 1885, by M. Hué, which says:

"We will see in the chapter which we dedicate to Russian petroleum that, thanks to the perfection of the apparatus invented, the problem is solved. The great fleet of steamers cruising in the Caspian sea, and going as far as the Volga; and the locomotives on all the Trans-Caucasian railways, are worked solely with petroleum for fuel. It is a matter of pride to us that the burning apparatus was invented by a Parisian; and this apparatus served as a model to the Russian engineers, giving the most splendid results.

"To apply this fuel to steamers, either merchantmen or men–of–war, would be to make a real revolution not only in steam navigation but also in the art of naval construction.

"We have seen steamers in Russia, such as the "Persia", one of the best on the Cunard Line, with space for 1400 tons of coal, until petroleum was adopted as fuel, after which only the fourth part of that space was occupied.

"Our big men-of-war of 6,000 tons require, at least, 1,000 tons of coal for a voyage of only ten days; whereas by using the same quantity of oil they might steam away for a month without touching at any port.

"Before the application of petroleum as fuel, 220,000 tons of oil were lost in Russia per annum, there being no employment to give it. To-day the steamers and locomotives of Caucasia use 500,000 tons, and those of the Trans-Caspian sea 280,000 tons.

"The petroleum burning apparatus is the invention of M. Sainte Claire Deville, perfected by Russian engineers in the year 1872; but, up till to-day, the war waged against petroleum by the great coal mining companies of France, England and Belgium has been so crude, and the struggle so tenacious, that oil has not yet been able to enter on the wide industrial field it is called upon to open. On the part of England, no quarter has been given in this war.

"It is a mistake to think that only the residue of petroleum—i. e., the remains after separating the volatile oils, parafines and gases—is applicable as fuel.

"The Russian sage Goulichambarof, whose opinion on this subject is law, asserts that there is no danger whatever in the use of crude petroleum as fuel after it has been exposed, for a time, to the air.

"Once crude petroleum has been exposed to the atmosphere, inflammable substances may be immersed in it, without the least danger of their catching fire. This is done almost constantly, by the inhabitants of Balakhani, in the petroleum lakes of that district.

"In summer, petroleum is soon relieved of its inflammable properties, and in less time in warm than in cold climates. The same result is obtained by submitting it to a high temperature, using the oil itself for fuel.

"The point of ignition of petroleum, on being taken from the wells, is 40° Celsius; and that of the residue 80° to 170; but the same crude petroleum which, on leaving the wells, ignites at a temperature of 40°, only ignites at 60° after being exposed for a few days, and after a few weeks' exposal, at 70° at least.

"During these last years, millions of tons of crude petroleum have been employed as fuel, in Russia, in all its different uses, and, so far, no noteworthy accident has happened.

"Up till to-day, there has not been a single explosion on the innumerable railways of Caucasia, where the locomotives, using petroleum as fuel, move, day and night, a large number of trains carrying the oil itself".

From the contents of this chapter, it will be seen:

$1^{st.}$ That the petroleum of the Department of Piura is superior to that of the United States.

$2^{nd.}$ That it has not the detestable odour of that oil.

$3^{rd.}$ That it has very little parafine, (a very inflammable substance).

$4^{th.}$ That the heavy oils are very valuable, being similar to those of Russia.

$5^{th.}$ That the kerosene extracted from it is of the best quality.

$6^{th.}$ That it is the most brilliant petroleum ever seen.

$7^{th.}$ That the crude oil may be employed as fuel, as is done in Russia, with no other treatment than that of exposing it to the atmosphere, for a few days, in open tanks.

## CHAPTER IV.

### *Markets for Consumption.*

The existence of petroleum in the Department of Piura having been demonstrated, not only by the authentic documents which we have published, but also by the fact of there existing, for some years past, in the petroleum fields of Máncora and Paita, two large establishments for refining kerosene, viz., Zorritos and the London Pacific Petroleum Company Limited, with a capital of nearly two million sols, the former turning out 6,000 cases of kerosene per month, and the latter distilling 300,000 litres of petroleum daily, with a thousand square miles of petroleum lands to dispose of, it now remains for us to inquire what would be the possible markets or rather let us say the markets of necessity, for the disposal of the product of the 2.500,000 wells which it is possible to drill in the 250,000 claims—calculating only ten wells to each claim— that is to say, taking as a basis the 1.000,000 hectareas of land which the engineer Prentice declares exist there.

The geographical position of the petroleum fields of Peru being lat. 3° 25' to 6° 5' South, and long. 81° 8' 4" to 3° 40' West, which is almost on the same parallel with Australia, China and other countries on the western border of the sea between them and Peru, it is to be presumed that these are naturally and necessarily the markets for Peruvian oils, whether exported in the form of crude petroleum or fuel, or in the form of refined oil or kerosene.

In this respect, the oils of Peru are so advantageously situated that they have nothing to fear from competition with the most favored country in the world.

It is known that the centre of the petroleum industry of the world is Pennsylvania, and that to place the oil in New York, the principal port for its exportation, it is run through 4000 miles of pipes, in order to save railroad freight.

This petroleum, which to-day monopolises the western market, leaves New York and, running the whole length of America, turns Cape Horn, and goes westward. They could be sent more easily and rapidly via the inter-oceanic railway between New York and San Francisco; but as there is no merchandise, however valuable it may be, that can stand the freight over that line's 3500 miles of road, it is clear that the only possible route for American oils is the dangerous route around Cape Horn, which voyage takes from 180 to 200 days to make; while the trip from the petroleum districts of Piura, Paita or Tumbez, to China, is made generally in 60 or 65 days, and economising 9000 miles of navigation, which of itself represents some pounds sterling.

The following table shows the advantages enjoyed in this respect by Peruvian over American oils.

COMPARATIVE Distances between New York and the Coast of Piura, and Western Ports.

|  | Miles. | Difference in favor o the coast of Piura |
|---|---|---|
| From New York to Madras | 11,745 | 2,135 |
| Calcutta | 12,330 | 2,630 |
| Singapore | 12,495 | 1,950 |
| Shangahi | 14,535 | 5,520 |
| Tokio-Japan | 15,165 | 7,035 |
| Sidney Australia | 13,050 | 6.000 |
| Auckland, N. Zealand | 14,025 | 8,145 |

*Real Distances.*

From the petroleum deposits, Mile

| | |
|---|---:|
| Piura, to Madras | 11,610 |
| „ New York to ditto | 11,745 |
| „ Piura to Calcutta | 11,700 |
| „ New York to do | 12,330 |
| „ Piura to Singapore | 10,545 |
| „ New York to do | 12,495 |
| „ Piura to Shanghai | 9,015 |
| „ New York to do | 14,535 |
| „ Piura to Tokio (Japan) | 8,130 |
| „ New York to do | 15,165 |
| „ Piura to Sydney (Australia) | 7,050 |
| „ New York to do | 13,040 |
| „ Piura to Auckland | 5,880 |
| „ New York to do | 14,025 |
| „ Piura to Hongkong | 9,226 |
| „ New York to do | 18,180 |

The only petroleum that could enter into serious competition with Peruvian petroleum would be that of California; but there the supply is not enough for the demand of that state alone, where, on the contrary, there is an annual deficit of a million barrels.

We may be asked: Which are the countries that could consume Peruvian petroleum, and what is the number of their inhabitants.

We answer:

Inhabitants.

| | |
|---|---:|
| English colonies | 209.000,000 |
| Chinese Empire | 403.000,000 |
| Japan | 36.000,000 |
| Siam | 5.750,000 |
| Australia | 3.411,000 |

| | |
|---|---|
| Colombia | 3.000,000 |
| Ecuador | 2.000,000 |
| Perú | 3.000,000 |
| Bolivia | 2.000,000 |
| Chili | 3.000,000 |
| Argentine Republic | 5.000,000 |
| Uruguay and Paraguay | 2.000,000 |
| Brazil | 12.000,000 |
| Total number of consumers | 689.161,000 |

The actual consumption, in the year 1889, of American petroleum in these countries, where it is introduced exclusively in the form of kerosene, was:

English gallons.

| | | |
|---|---|---|
| Australia | | 5.173,801 |
| Aden | 270,000 | |
| Burmah | 2.420,570 | |
| India | 22.458,840 | |
| Penang | 607,600 | |
| Singapore | 1.676,760 | 27.433,770 |
| Siam | | 580,680 |
| China | | 16.836,764 |
| Japan | | 9.424,560 |
| Java | 16.578,380 | |
| Maccassar | 256,000 | |
| Sumatra | 705,000 | 17.549,380 |

*South America.*

| | |
|---|---|
| Colombia | 2.468,324 |
| Ecuador | 75,170 |
| Chili | 1.755,530 |

| | | |
|---|---:|---:|
| Hawai | 164,640 | |
| Bolivia | 176,420 | |
| Perú | 335,592 | |
| Argentine | 5.000,000 | |
| Uruguay | 446,937 | |
| Brazil | 9.000,000 | 19.422,713 |
| Total number of gallons | | 96.421,668 |

equivalent more or less to 400,000 English tons, or say 7.000.000 cases, which, at S/ 2 each, represent the sum of 14.000,000 sols.

It is possible that Peruvian Petroleum, on account of the facility of its extraction, its abundance and good quality, and its being found on the coast, may some day compete with North American kerosene, in the most thickly populated centres of Europe, if the Panamá canal is completed.

It remains for us simply to consider the value of petroleum as fuel.

According to some rather deficient data before us, the amount of English coal consumed in South America is as follows:

| | Tons. |
|---|---:|
| Colombia | 136,000 |
| Ecuador | 117,000 |
| Perú | 110.000 |
| Chili | 600,000 |
| Argentine Rep | 800,000 |
| Uruguay | 338,000 |
| Brazil | 1.240,000 |
| Total | 3.411,000 |

This shows that South America, which would be the nearest market for the consumption of Peruvian petroleum, as combustible, consumes 3,411,000 tons of English coal per annum and which, calculated at the average price of £ 3, gives the enormous sum of £ 10.233,090, or 51.165.000 sols.

By applying crude petroleum to the same uses as coal, which would represent an economy of 50 per cent., it is clear that the half of the value of the coal, or S/ 25.582,500 per annum, would be clear profit to the consumers, and that that sum would represent the extent of the industry in Perú.

In order to give our readers an approximate idea of this powerful industry in the United States, we submit the following table:

UNITED STATES.

*Production of Petroleum.*

| Years. | Barrels. |
|---|---|
| 1860 | 500,000 |
| 1861 | 2.113,609 |
| 1862 | 3.056,690 |
| 1863 | 2.611,309 |
| 1864 | 2.116,109 |
| 1865 | 2.497,700 |
| 1866 | 3.597,700 |
| 1867 | 3.347,300 |
| 1868 | 3.646,117 |
| 1869 | 4.215,000 |
| 1870 | 5.260,745 |
| 1871 | 5.205,341 |
| 1872 | 5.890,248 |
| Carried forward | 44.057,868 |

|   |   |
|---|---|
| Brought forward | 44.057,868 |
| 1873 | 9.890,964 |
| 1874 | 10.809,852 |
| 1875 | 8,787,506 |
| 1876 | 8.968,906 |
| 1877 | 13.135,771 |
| 1878 | 15.163,462 |
| 1879 | 20.041,581 |
| 1880 | 26.032,421 |
| 1881 | 29.674,458 |
| 1882 | 31.789,190 |
| 1883 | 24.385,966 |
| 1884 | 23.596,945 |
| Total | 266.334,890 |

This represents the value of $443.991,980 American gold.

### EXPORTS OF PURIFIED PETROLEUM.

*Kerosene of the United States.*

| Years. | Gallons. |
|---|---|
| 1864 | 23.210,369 |
| 1865 | 25.496,849 |
| 1866 | 50.987,341 |
| 1867 | 70,255,481 |
| 1868 | 79.456,888 |
| 1869 | 100.636,684 |
| 1870 | 113.735,394 |
| 1871 | 149.892,691 |
| 1872 | 145.171,583 |
| 1873 | 187.815,187 |
| Carried forward | 946.658,467 |

|  |  |
|---|---:|
| Brought forward | 946.658,467 |
| 1874 | 247.806,483 |
| 1875 | 221.955,308 |
| 1876 | 243.660,152 |
| 1877 | 309.198,914 |
| 1878 | 338.841,303 |
| 1879 | 378.310,010 |
| 1880 | 433.964,699 |
| 1881 | 397.660,262 |
| 1882 | 559.954,590 |
| 1883 | 505.931,622 |
| 1884 | 513.660,092 |
| 1885 | 574.668,180 |
| 1886 | 577.781,752 |
| Total | 6.240,051,834 |

representing a value of $ 847.126,550 American gold.

The 200 petroleum refineries working in Bakhu (Russia) have produced the following:

| Years. | Tons. |
|---|---:|
| 1872 | 16,400 |
| 1873 | 24,500 |
| 1874 | 23,600 |
| 1875 | 32,600 |
| 1876 | 52,100 |
| 1877 | 72,600 |
| 1878 | 97,500 |
| 1879 | 110,000 |
| 1880 | 150.000 |
| 1881 | 183,000 |
| 1882 | 202,000 |
| 1883 | 206,000 |
| Total for twelve years | 1.180,300 |

The output for the years 1883 to 1885 may be analysed as follows:

*Gallons of kerosene.*

| 1885 | 1884 | 1883 |
|---|---|---|
| 137.000,000 | 109.000,000 | 72 000,000 |

*Crude Oil.*

| | | |
|---|---|---|
| 19.000,000 | 10.000,000 | 10.000,000 |

*Residue.*

| | | |
|---|---|---|
| 170.000,000 | 142.800,000 | 87.000,000 |

*Lubricating Oil.*

| | | |
|---|---|---|
| 7.950,000 | 7.200,000 | 5.000,000 |

*Benzine.*

| | | |
|---|---|---|
| 140,000 | 380,100 | 240,000 |
| 334.090,000 | 269.380,100 | 174.240,000 |

Table showing the total annual output of kerosene in the world:

### 1885

*North America.*

Barrels of 160 lbs.

| | |
|---|---|
| Canadá, 200 wells ........................ | 900,000 |
| United States, 25,000 wells............... | 64.235,081 |

*South America.*

Trinidad....................... (not worked)
Venezuela ............................ ditto

| | |
|---|---|
| Carried forward................... | 65,135.081 |

|  |  |
|---|---|
| Brought forward............... | 65.135,081 |
| Perú ............................................. | 300,000 |
| Bolivia........................(not worked) | |
| Argentine..............................ditto | |

### Australia.

|  |  |
|---|---|
| New Zealand..................(not worked) | |
| Australia ........................................ | 80,000 |

### Asia.

|  |  |
|---|---|
| Japan, 200 wells............................. | 35,143 |
| China................................ (Unknown) | |
| Birmania......................................... | 1.000,000 |
| India................................(not worked) | |
| Trans-Caspian country, 1 well........... | 11 6,250 |
| Bakhu, 400 wells............................. | 15.62 5,000 |
| Caucasia, 250 wells......................... | 50,000 |

### Europe.

|  |  |
|---|---|
| Roumania, 1,200 wells...................... | 125,300 |
| Galicia, number of wells unknown........ | 5.00 0,000 |
| Germany, 200 wells.......................... | 300,000 |
| Italy..............................(not worked) | |
| France ..................................ditto.. | |
| Total number of barrels............... | 87.766,77 4 |

Fluctuations in the prices of petroleum, in the United States, for the following years.

| Years. | Maximun. | Minimun. | Average. |
|---|---|---|---|
| 1860 | 19.25 | 2.75 | 9.59 |
| 1861 | 1.00 | 10 | 49 |
| 1862 | 2.25 | 10 | 1.05 |
| 1863 | 3.95 | 2.25 | 3.15 |

| | | | |
|---|---|---|---|
| 1864 | 12.12 | 4 | 8.06 |
| 1865 | 8.25 | 4.62½ | 6.59 |
| 1866 | 4.50 | 2.12½ | 3.74 |
| 1867 | 3.55 | 1.75 | 2.41 |
| 1868 | 5.12 | 1.95 | 3.62½ |
| 1869 | 6.95 | 4.95 | 5.63¾ |
| 1870 | 4.52½ | 3.15 | 3.84 |
| 1871 | 4.82½ | 3.82½ | 4.34 |
| 1872 | 4.02½ | 3.15 | 3.64 |
| 1873 | 2.60 | 1.00 | 1.83 |
| 1874 | 1.90 | 69 | 1.17 |
| 1875 | 1.75 | 1.03 | 1.35 |
| 1876 | 3.81 | 1.80 | 2.56½ |
| 1877 | 3.53¼ | 1.80 | 2.42 |
| 1878 | 1.65¼ | 82⅛ | 1.19 |
| 1879 | 1.18⅛ | 67⅛ | 85⅞ |
| 1880 | 1.06¼ | 78 | 94¾ |
| 1881 | 94½ | 76⅞ | 87⅞ |
| 1882 | 1.14 | 54⅜ | 78½ |
| 1883 | 1.24 | 83¾ | 1.04¼ |
| 1884 | 1.11 | 63½ | 1.02¾ |

## CHAPTER V.

*Its Future as Fuel.*

The movement which has lately been going on in Europe in favor of petroleum as fuel, notwithstanding the fact that Europe is the great centre of coal mining interests, (the mines of France, England, Italy and Belgium giving daily occupation to thousands of workmen, and being worked by old companies, with capital representing many millions of pounds sterling) proves either that the supply of coal is being exhausted, and its extraction becoming dearer every day, or that liquid fuel possesses such advantages over its rival of so many centuries that, after a hard battle, the defeat of the lord of the entire globe, up till to-day, is evidently being achieved.

In this respect, we must point out certain facts which are being realized, all of which are of the greatest importance.

The fact that in the French legislature and the English parliament, according to the statement of Captain Coxmichael, the question was discussed of replacing coal with petroleum in the navies of those countries (of the definite results of which we are entirely ignorant), after having tried liquid fuel a number of times in the port of Cherbourg, on board the *Puebla*, and in the steamers running between Paris and Rouen, where it gave every satisfaction; and, above all, after using it widely for all domestic purposes, in the great factories and the mills of Paris during the siege of 1871; all goes to show that the trials in France have given good results.

Similar progress manifests itself in England, notwithstanding the lentitude with which all innovations are received in that country, as the participation of ther Bitish Admiralty is being felt, and it is represented by its eminent engineers, at all the actual trials and conferences connected with liquid fuel, which are frequently repeated on the Thames, where steamers run to-day using the new combustible.

Besides these facts, we learn from Mr. Henwood, who took part in the conference in London on March 13th last, that steps were being taken to store petroleum in different parts of the world, in order that naval squadrons, as well as merchantmen, could obtain it easily.

Mr. James Holden, of the Great Eastern Railway, declared, at the same conference, that he had used petroleum in many of the locomotives on that line, and that one of them, prepared for the purpose, had already run 47,000 miles, over the road between London and Norwich, without the slightest mishap or difficulty. These long experiments, which have lasted years perhaps, have lately, at all events every thing points that way, given every satisfaction.

To these certainly very significant antecedents we may add, according to what we have been assured, that the Italian government, following the example of Russia, after the last experiments made in their dockyards at Spezia, have decided to use liquid fuel in their big navy.

For the moment, the great obstacle to adopting this combustible definitely consists undoubtedly in being able to obtain it constantly and at a fair price.

In this respect, Peruvian petroleum is called upon to play the most important role in the commerce of the world.

The petroleum of Bakhu, the old Persian city, in order to reach European markets, would have to run over an immense net-work of railways, hundreds of miles long, the freight over which the oil could not support. On the other route, which is the longest, but the cheapest, there is the inconvenience of transhipment; for, in order to arrive at European centres of consumption, the oil would naturally have to traverse the Caspian, Baltic and Mediterranean seas, besides a few railways between them, and finally go through the Dardanelles, a fortified Turkish pass, on the free transit through which it could not always count, and much less so in case of war.

Apart from the fact that for Russian petroleum to reach Europe it would be burdened with heavy railway and steamer freights and duties of all kinds, we do not think that any European nation, however badly off it might be for fuel, would submit itself voluntarily to the service of Russia, for an article of prime necessity, as fuel is, nor submit its navy to the tutorship of such a powerful nation.

For this and other reasons, of no less importance, Russian petroleum, in the form of fuel, will never be able to take a firm hold on European markets, although it may do so in the form of illuminating oil or kerosene.

Canadian petroleum, which is another important source of fuel, and which could advantageously supply European markets, is not abundant, as the 3000 barrels produced daily at Oil-Springs, Wyoonning, Petrolia and Eniskillen, all very far from the coast, barely suffice for home consumption.

The petroleum of Pennsylvania, in the United States, is the only oil that reaches Europe under fair economic conditions, and it has, as a matter of fact, the monopoly, for illuminating purposes, in the four corners of the earth; but this is always in the form of kerosene and never as fuel; and it will always be thus with the North American product, as, exported in this form, it leaves splendid profits, while the residue is used at home, without paying freights or duties.

The supplies of petroleum in Bolivia and the Argentine Republic are not available, as they are found, in both countries, in the roughest regions, hundreds of leagues from the coast, and in semi-savage localities.

The exceptionally favorable geographical position of the Department of Piura makes it accessible, beyond dispute, to all the markets of the world.

The petroleum fuel of Peru, which, as we have explained in the preceding chapter, may easily monopolise, with many advantages to consumers, all the markets of South America, as well as the centres of consumption of the nations of the west, may also, some day, and without experiencing any great difficulties, make its appearance at a central point on the Atlantic ocean, and, once there, reach the markets of Europe, Asia and Africa.

As the petroleum of Pensylvania, in order to arrive at its port of exportation, which is New York, is run through an immense net-work of iron pipes 4,000 miles long, uniting the 25,000 wells that are being worked there; and as petroleum in Bakhu is taken to the centre of Persia by similar iron pipes, but only 1,500 miles long (the property of the house of Noble Brothers, the discoverers of dynamite), crossing rough land, chasms, big rivers, seas, lakes and mountains of ice; we see no reason of any kind why the combustible oils of Peru should not reach Panama and the Atlantic port of Colon, by such an easy and economic system as the iron tubing, like that of Pennsylvania and Russia. It must be remembered that the distance from Sechura, Paita or Tumbez, which are the three petroleum districts of Piura, to Colon, is barely 900 miles, which is a mere nothing to the power of capital.

We have two specimens, on our coast, of these great works, although on a small scale: one is the pipe that takes water from Arequipa to the port of Mollendo, over cordilleras and high hills, one point being 9,000 feet above the level of the sea, and which is 120 miles long; and the other the pipe taking water from Pica to the port of Iquique, under no less significant engineering difficulties than the former, and which is of even greater length.

Until the great work indicated is carried out by bold and enterprising men, which is what is wanted in such cases, the fuel could be taken in large tank steamers, from the coast of Piura to Panama—a great centre of consumption for many steamship lines—and from there to Colon, through a pipe fifty-four miles long. From this point, which we may call central, all naval squadrons and merchantmen might be supplied, and vast deposits erected even in most out-of-the-way places.

This is how the petroleum of Peru, with the aid of capital, may make its appearance on the Atlantic, and be easily transported to Europa, Asia and Africa, as economically as across the Pacific to Australia, Japan and China, as we have already demonstrated.

## CHAPTER VI.

### *Works already Established.*

Two large establishments have been founded, up till to-day, in that extensive country for refining petroleum and producing kerosene, the greater part of which is consumed, in the latter form, in Ecuador, Peru, Bolivia and Chili.

Signor Faustino G. Piaggio has the honor of being the first man to found a first class establishment of the kind in South America; and the first also to defy the competition of North American kerosene, which, until a short time ago, monopolised this market.

The kerosene of Zorritos, as Signor Piaggio's refinery is called, has received premiums for its excellent qualities at several European exhibitions, and obtained the gold medal at the South American Exhibition in Berlin, in 1884.

The petroleum supply of Zorritos is situated on the sea shore, in the hacienda of Máncora, in the province of Tumbez, about 34 kilometres to the south of that city; and, according to all the data in our possession, it is the richest petroleum district in the North.

This establishment for obtaining kerosene and other products of petroleum, was, as we have said, the first one founded in South America. It has all the modern apparatus required for working on a vast scale.

The Minister of Finance, Don Eulogio Delgado, in his last memorial to Congress, describes this place as follows:

"These works have 54 claims of 40,000 square metres each, constituting an area of 2,160,000 superficial metres. It covers the ground from the point of Malpaso Grande to the Bocapan gulley, and, stretching inwards from the sea shore to the hills of the province of Tumbez, is situated at 3° 41' South lat., and 80° 37' long. west of Greenwich.

"Eleven wells are being worked with pumps necessary to extract the quantity of petroleum required to fill the demand for the refinery established there, which has facilities for refining 6,000 cases per month.

"The works include blacksmith, tinsmith, machine carpenter and cooper shops, and the respective stores for materials, tools and different apparatus.

"There are also dwellings for the employees and laborers, and for the port authorities.

"The establishment comprises in all 25 buildings, without counting the small village of Sechurita, which has been laid out two kilometres away from the offices, and is the residence of the greater part of the laborers and their families.

"The works are connected with each other by railroad (the Decoville system), and the principal centres communicate with the Manager's Office by telephone.

"There are, in fine, machines for all kinds of work, and a good proportion of tanks, pipes and other apparatus required for the same.

"A large sum of money was invested in this establishment by its primitive owners, and no small one by the present manager and propietor, Signor Piaggio, an industrious and respected merchant of Callao.

"Four hundred workmen are employed in Zorritos, and the kerosene for illuminating the streets of Tumbez is supplied gratis by that factory.

"This establishment, with more capital and active mercantile operations, is called upon to be another great centre for kerosene and other products of petroleum; that is, when it has capital sufficient to transport the oil in its own steamers, and thus become independent of the established lines, which limit its exportation on account of the few trips they make".

The other establishment for refining petroleum, following Zorritos in chronological order, is that of the London & Pacific Petroleum Company, Limited, known by the names of Pariñas or Talara, which is its port.

It is situated in the centre of the petroleum fields of Piura, in the province of Paita, 52 miles from that port, in the district of Amotape, between the Chira and Túmbez rivers, occupying a large part of the hacienda of Pariñas, known by the name of Brea, the property formerly of Don Genaro Helguero, who sold it to Herbert W. C. Tweddle, Esq., for £20,000, in 1887.

The petroleum grounds of this company comprise an extension, it is declared, of 1,000 square miles; but it has only registered ten claims, of 40,000 square metres each, on which the proprietors pay contributions to the Government.

Mr. Tweddle, an intelligent and hard-working man, of extraordinary industrial conceptions, is the same person who, in 1875, proposed to the Russian government to transport, for his own account, petroleum from Bakhu to Poti, by means of an immense network of pipes, on the basis of the concession to him of a million acres of land, and the monopoly of the transportation of the whole output for forty years. The technological society of St. Petersburgh was consulted, and it opposed the scheme, basing its decision on the grouud that before the forty years were up, the income derived from the land would amount to 420.000,000 francs. In the meantime Peru owes to Mr. Tweddle the first and largest establishment for refining petroleum and supplying oil fuel, thus contributing towards the fixing of the bases for the new industry, which at no distant time will take its rapid and long expected impetus.

In order to avoid repetitions, we will quote the official description of the Minister of Finance, and which is contained in his report. It is as follows:

TALARA.

*Petroleum wells.*

It was not in vain that I last year called the attention of Congress to the subject of petroleum wells, in the provinces of Paita and Túmbez, as sources of great wealth for the country's future. To-day I am glad to announce that that industry has been initiated with unexpected enthusiasm, and that, in the natural course of things, it will be the forerunner of others of the same and other natures.

To convey an idea of what has been done in this direction, allow me to give you a brief description of the companies established there.

The London & Pacific Petroleum Co., with the object of competing, at all events on this side of the continent, with the two great companies of the world, the Noble Co., in Russia, and the Standard Oil Trust Co., in the United States, has been organised with a capital of £ 250,000 for the purpose of establishing itself in Talara, also called Brea, a port on our coast, between Paita and Tumbez, where it has found that vast field for its development which the demand for the article requires.

In Talara, one of the best ports on our coast, on account of the smoothness and depth of the sea at that point, works have been carried out, with surprising rapidity, which to-day comprise the following:

In Negritos, nine wells which give 250,000 to 300,000 litres of petroleum per day.

A powerful pump to force the petroleum through a pipe 11 kilometres long, to the tanks of the Talara refinery. This pump has sufficient force to supply three times the quantity mentioned, or one thousand tons of crude oil daily.

Two stills for the production of kerosene and capable of distilling 120,000 litres per day. Another still, of double capacity, is in course of construction, and there is material on hand for the making of a greater number, because, before the end of the year, it is expected that the new wells will yield double the quantity of petroleum, that is to say, more than a million litres daily.

They have put up five iron tanks, with a joint capacity for the deposit of more than half a million litres of oil, and others are in course of construction, with a capacity of half a million litres each, destined not only for the supply of the factory, but intended also for one of the principal ports on our coast, and to supply, at all times, the demand for petroleum, kerosene and lubricating oils.

They have erected a good number of houses for workmen, and a stone building, 275 metres long, in which they have installed the blacksmith, machine and tinsmith shops, which latter can turn out five thousand tins daily. There is another stone building adjoining the foregoing, for the manufacture of three thousand cases per day.

The tins are filled in a separate building, by a special apparatus in the boiler shop, where stills and tanks are made, and rivetted by means of compressed air.

They have built a lighthouse in the port, for the use of their vessels, leaving the Government to collect the dues on other ships; and the mole in course of construction, to which six steamers may be moored at a time, will insure the convenience and safety of passengers.

The port of Talara, which, when the works were first begun, had no other drinking water to rely upon than what was brought from a place thirty kilometres distant, by beasts of burden, has now running water, raised from wells by windmills, and conducted through pipes 7 kilometres long, and which not only fills all the wants of the establishment, but is also used in the raising of vegetables.

A tank steamer, one of the five which have been ordered from England, has already arrived. Their object is to supply petroleum, kerosene and oil to the tanks on the coast, thus enabling the public to economise the extra expense of tins and cases, and to reap the benefit of the saving made by unloading simply from tank to tank with hose.

The Company will continue to extend the works, with which object they will augment their capital to £ 500,000, as, from what I hear, they intend to supply the Japanese and Chinese markets.

As will be seen, the Company have begun and will continue this industry, on a large scale, basing their calculations on the profits obtained by large outputs.

The kerosene elaborated by them appears to be of as good, if not of better quality as that imported in considerable quantity from the United States.

Besides having lowered the price of the article in this market to such a point that only the native product is consumed, the importation of coal is also being diminished, as the employment of petroleum, has begun for fuel for locomotives; the caloric power of 35 kilos of petroleum, according to experiments made, being equivalent to 90 kilos of coal, but, as it is a little dearer, the economy in its use is fifty per cent.

In Russia and other places, this new combustible has already been employed for locomotives and we should congratulate ourselves on the fact that, within the limits of our own territory, we have a relatively cheap combustible, which we so badly needed before, and which we were formerly obliged to import.

From what I can see, we will soon have petroleum works here, as good and extensive as the largest in the world, with the advantage of being completely independent, established as they will be on the sea shore, with all the elements necessary for the manufacture of acids, with which object the respective pipes are being laid down.

The manufacture of acids will give birth to a number of industries in the country, and which could not be undertaken without counting upon the supply of acids at a moderate price:

We have, therefore, a valuable industry, the mother of many other important ones, besides that indispensable element, fuel, advantageously situated for exportation to the coasts of the Pacific; on which account, capital will come here from abroad to augment the wealth of the country, and give impulse to new and, at present, latent industries, which are only waiting to be established to yield most abundant remuneration.

As will be seen, the districts of Negritos, Brea and Zorritos guarantee the result of the other undertakings which will be established in the country at no far distant time.

## CHAPTER VI.

*Fields which are being Explored.*

### THE PETROLEUM REGION OF THE SOUTH.

The petroleum fields in the South, comprehended in the extensive district known as the peninsula of Punta de Aguja, in the district of Sechura, in the province of Piura, are, at this moment, an object of careful study, on the part of the North American engineer Major Taggart and his associates, with the most perfect apparatus and machinery known.

These petroleum fields, discovered only last year (1890), are situated on the sea shore, at distances varying from 400 to 16,000 metres from the water's edge, south of the port of Sechura.

These vast petroleum beds lie between long. 80° 58' and 81° 11' West of Greenwich. and lat. 5° 48" to 6°.10" South, 17 leagues, or 51 miles, distant from Sechura—an active agricultural and commercial centre, of 9,095 inhabitants—and 60 miles by sea from the first class port of Paita.

The whole district is crossed by a range of hills, lying very near to the coast, and which begin at Punta Pisura, the bay of Sechura, stretching in a southeasterly and afterwards an easterly direction, and end in the desert of Sechura, near the salt deposits of that name. These hills rise as high as 400 metres, are covered with pasture in the summer, and are favored with an abundant rainfall.

A number of small *cañons* begin at the centre of the hills, and are prolonged as far as the sea, covered with luxurious vegetation, the most important of them being the Avisp and Ramazon, as they have a few water wells, the water of which, although saltish, is good for animals.

The level portion of the district, on the desert side, is covered, at some points, with a thick layer of earth mixed with nitrate, which only contains small quantities of nitrate of soda, and is composed principally of layers of ferruginous clay, alumn, chalk and salt, which latter article is inexhaustible in the extensive salt deposits of Sechura, covering many square leagues, and extending thirty kilometres away from the haven of the salt works.

At the end of Punta Falsa or Avisp *cañon*, the cliffs 90 metres above the level of the sea, descend rapidly, and a plain is formed, fifteen kilometres long, rather more than less, by two kilometres wide, which stretches between the sea and the hills.

This extensive plain is prolonged as far as the Ramazon cañon, where it ends at the cliff, when the land again rises, and forms another plain higher than the other. This "bank" is formed of extensive layers of sulphur, which may be utilised for the manufacture of sulphuric acid, so necessary for refining kerosene.

It is on these two plains that the petroleum districts of Garita and Reventazon lie, at a distance of from 430 to 3,000 metres from the sea, and which are perhaps the richest of the petroleum regions of the South.

The first plain, Garita, is a depression of the ground, full, according to the opinion of experts, of petroleum, which overflows the surface, in many places, mixed with water. The soil of this place is completely impregnated, or saturated with oil, and the oil is struck at only two metres from the surface.

In the second plain, Reventazon, it has happened that the subterranean gases of the petroleum have upheaved the ground here and there, and the petroleum has issued from beneath it, spreading over the surface, from the centre to the circumference, forming large mounds, which are nothing but petroleum hardened by atmospheric action. Some of the mounds have a circunference of 200 metres, and are 10 or more metres high.

These miniature volcanoes, especially those near the hills, and which are known in Piura by the name of "stone tar" [*piedra brea*], are in a constant state of activity. On the sides of some of the latter, fissures are observed, from which petroleum issues, mixed with salt water.

The temperature in the excavations made so far is rather high, and the odor escaping from them is that of kerosene.

To the West of these petroleum fields, there must be large subterraneum streams of petroleum, which are lost in the sea, as the surface of the water is covered with large patches of oil, reflecting all the colors of the rainbow.

These petroleum beds have several available ports, but the most preferable of all is the magnificent bay of Nonura, where there are also vast petroleum fields, and which is about nine miles from Garita and Reventazon.

At the foot of the northern beds of the region in question, and at the extremity of Pisura-point, the coast turns to the Northeast, and the bend thus formed is the beautiful and sheltered bay of Sechura, 24 miles long, and 18 miles from Pisura-point. Sheltered from southerly and westerly winds, it is very safe, while its waters are continually in a state of calm, and it can harbour hundreds of vessels.

In the centre of the bay, about six miles from Pisura-point, there is a beautiful and extensive plain, some leagues long, covered with a thick growth of smooth-podded tare, known by the name of Bayobal woods, where the centre of the petroleum refineries may be established, when the oil can be taken there through pipes.

The petroleum region of the South, or Sechura, not only comprises the district described, but extends as far as the haven of the salt works, comprehending the points and bays of Nonura, Pisura, Bichayo, etc., etc.

In this region, liquid petroleum is found at from one metre and upwards from the surface, with slight variations.

A syndicate formed in London is having the petroleum fields we have just described, studied, for its own account.

## Region of the Centre.

This comprises a part of the province of Paita and the port of that name, and extends from the Máncora gulley to the South of the Silla de Paita and Foca-point.

These fields, which include a great extent of petroleum country, constitute Negritos, Brea, Talara, Siches and Organos, the first mentioned places being thoroughly worked.

Petroleum is visible at very few points in this vast territory, and yet it appears to be very abundant there.

The Region of the Centre, apart from the beds mentioned, includes another district, extending for 1·7 leagues, from Lobo-point to the banks of the Chira river, and is bounded on the North by Brea, the area of which, calculated by a German engineer, is 11.310,000,000 square metres, comprising Lobo-point, Tortuga-point, Playa Perico, Caleta del Horno, Playa Grande, Playa Paquié, Pescadores, Loradan-point, Chué-point, Tablazo de Paita, Panteon Ingles, Sorá-point, Colan, Tablazo de Cabo Blanco, etc.

The petroleum supply in this district does not show itself excepting in Panteon Ingles, Sorá-point, Tutate, Pescadores, Lobo, Tortuga, Paquié and Colan; but extensive layers of solidified petroleum are found underneath the sand which covers them. The oil does not show on the surface, doubtless on account of the great height of the ground above the level of the sea, and the narrow width of the coast here, or on account of the greater depth at which the supply lies. At Playa Pescadores, there is a magnificent port of the same name, and the bay of Caleta del Horno, between Playa Chica and Playa

Grande, where the oil might be embarked without any necessity for taking it to Paita. In the Paquié gulley, there are fresh water springs sufficient to supply this whole district. It is situated at long. 81° West, and lat. 5° to 6° South.

The climate is unsurpassable: dry, breezy and very healthy. The heat is not excessive, as there is a sea breeze blowing constantly. There is a great abundance of fish of all kinds, better, perhaps, than those along the remainder of the coast of Peru. In addition, all kinds of provisions, fruits, etc., may be produced here.

Coasting steamers pass close by constantly, and could easily touch at any of the ports or bays.

No serious study has ever been made of this region, and we regret this, as it is the one that has most resources, and the greatest number of inhabitants, towns and ports.

## The Region of the North.

This comprises the whole of the province of Tumbez, from the Máncora gulley to the Zarumilla river, which divides Peru from Ecuador, its extent being, in all, a hundred and fifty miles.

The only part of this district which has been studied, and that barely enough, is Máncora, the survey of which will be continued throughout the present year, by three different companies, formed for the purpose, and which will soon send engineers and apparatus to begin work.

Up till to-day, no study has been made of the country north of the Tumbez river, the explorers limiting their operations to taking advantage of the land where petroleum showed itself on the surface, either in the form of mounds, as in Sechura, or in the liquid form, in which it appears spontaneously at the surface, as it does nearly all over Máncora.

The district of Máncora, which is rich in petroleum fields such as Zorritos, and the beds recently acquired by Don Manuel A. de la Lama and Nephews, which comprise a vast extent of territory, and those of Mendizabal Hermanos, are called upon to be an industrial centre of the first order. It is divided into three little branches, belonging to the primitive owner Don Diego de Lama.

The petroleum district of which we now speak begins in the gulley of Charán, ten miles to the South of the Tumbez river, and, extending along the coast to the gulley South of Cape Blanco, comprises an oil field of approximately forty leagues long, by an average width of twenty leagues, from the low hills of the cordillera to the sea.

The coast of this region is full of sinuosities; but its general direction is southwest, beginning at Malpelo-point, which forms the southern end of Tumbez bay. In the little offing at La Cruz, near Charán, there is a point stretching out into the sea, and another at Malpaso Grande, which forms, in Malpaso Chico, a merely mediocre anchorage ground, for although pretty deep, it is exposed to southwesterly, and, what is worse, to northerly winds. From Malpelo-point, as far as the Charán gulley, the coast is low, distinctly separating the district from the valley of Tumbez.

La Punta, which follows it, forms the northern, and Zorritos-point the southern limit, of the bay of that name, which is an excellent anchorage for vessels, and where thirty feet of water are found, at low tide, a thousand feet from shore.

Further to the South, the nearest place of importance is Punta de los Picos, South of Bocapan, which is another and very much frequented port, with an excellent anchorage.

Immediately to the North of Cape Blanco is Cape Máncora, where there is a very important bay, well protected and close to Caña Dulce, a district very rich in petroleum and asphalt.

The hills in this region begin very near the shore, and rise at the rate of about a hundred feet per mile.

The surface of the ground is covered with different kinds of clay, sand and fossiliferous conglomerates, composed of gravel, stones and round rocks, held together with carbonate of lime, sand and ferruginous earth.

The petroleum stained rocks of this region appear to be sandstone containing some chloride of sodium, the greater part of them being saturated with oil, and the brittle layers of which are so weakly united that they allow the petroleum to escape through the cracks. The oil which inundates the beach may be seen dropping from the rocks exposed to the constant action of the sea.

This abundance of petroleum is noticed from Malpaso Chico onwards, and, at intervals, it is found as far as Cape Blanco, i. e., a distance of thirty leagues or ninety miles.

The district where stained rocks are found is about twenty leagues long by seven wide, from Malpaso to Caña Dulce, making, in all, 806,400 acres.

The northern petroleum fields, in the province of Tumbez, have large and sheltered ports, where the oil may be shipped, with great facility, by means of iron piping and hose, through which it could be run and poured into tanks on board ship. The largest vessels may anchor in any of these ports.

The temperature of this region is, as a rule, agreeable. Lower than that of Guayaquil and Tumbez, the thermometre seldom rises to 85° F.; and is generally at 59° to 63° at night, and 67° to 71° during the day.

Although so close to Ecuador, the sea-breezes which blow refreshingly from the southeast during ten months of the year, and from the Northwest during the other two, sensibly moderate the atmosphere and make the climate truly healthy.

The points where petroleum shows itself are: Pan viejo, Zorritos, Sechurita, La Cruz. La Garita, Malpaso, Q. (*quebrada* or gulley) Tucillal, Tres puntas, Q. Heath, Corrales, Q. Boca de Pan, Piedra redonda, Punta Sal, Rivera de Tumbez, Salado de Averías, Punta de Mero, Piedras redondas, Canoas, Punta de Pico, Peña negra, Q. Culebras, Barranco de Peroles, Cerro blanco, Q. Seca, Sal chico, Palo santo, Sal grande, Q. Gigantal, Cardalito, El bravo, Boca pan, Q. del pozo, Q. Peroles, Q. Tijeritas, Salado de los preciados, Salado de realengo, Salado de azotadero, Cuesta de averías, Q. Hucco, Mal pelo, Mal paso, Pedro viejo, Boca de Tumbez, El hospital, El abejal, Orillas de Tumbez, Estero de taical, Punta de bestin, Punta de ñuro, Q. de ñuro, Q. verde, Q. Beintia, Tablazo de cabo blanco, La Breita, Siches, Organos.

The number claims of which the Government have thus far either given possession, or, to obtain which, the necessary legal formalities are being complied with, does not exceed twenty two hundred, and are distributed as follows:

|  | Claims. |
|---|---|
| Region of the Centre (Paita) | 1000 |
| Region of the North (Tumbez) | 600 |
| Region of the South (Sechura) | 600 |
| Total number of claims | 2200 |

or 8,800 hectares, the equivalent of 22,000 acres, or 88,000,000 square metres, which bring in to the national treasury, in the form of contribution of mines, at the rate of S/. 30 per annum for each claim, the sum of S/. 66,000, which is the largest sum known yielded by such a limited space in the country.

## CHAPTER VII.

### *General Aspect of the District.*

The petroleum deposits of Peru are not situated in a desert, but in the commercial and productive centre of Piura, the extreme northern Department of the Republic.

The area of this vast portion of our territory comprises 70,463 square kilometres, divided into five provinces, those on the coast being Piura, Paita and Tumbez, and those in the interior Huancabamba and Ayabaca. The total population of 135,615 is as follows:

| Provinces. | Men. | Women. | Total. |
| --- | --- | --- | --- |
| Piura | 26,824 | 28,275 | 55,099 |
| Tumbez | 2,940 | 2,938 | 5,878 |
| Paita | 10,168 | 10,909 | 21,077 |
| Ayabaca | 17,411 | 18,165 | 35,576 |
| Huancabamba | 8,446 | 9,539 | 17,985 |
| | 65,789 | 69,826 | 135,615 |

The different races which make up the inhabitants are the same as those which compose the population of all Spanish America, Europeans being established chiefly in cities like Piura, Paita, Huancabamba and Ayabaca.

As a general rule, the inhabitants of this department are of robust physique.

Three large rivers run through this extensive territory, from the mountains of the interior to the sea. They are the Chira and the Tumbez, navigable, respectively, all the year round, for 200 miles and 60 miles.

More than 300 haciendas lie on the banks of these rivers, the principal products of which are maize and cotton of superior quality.

All the products of the district are of the torrid zone, and the crops are reaped as follows:

Sugar-cane .................... 4 crops per year.
Cotton ............................ 2 ,,
Maize ............................. 4 ,,
Rice ................................ 3 ,,
Ramié ............................. 4 ,,
Tobacco .................... 5 to 8 ,,
Coffee ............................. 2 ,,
Hemp ............................. 2 ,,
Cacao, the whole year round.
Potatoes ......................... 2 ,,
Sweet potatoes .............. 3 ,,
Yucas ............................. 2 ,,
Beans ............................. 4 ,,
Cocoanuts, the whole year round.

Almonds, walnuts, chestnuts, aniseed, wheat, barley, pumkins, chick-peas, Lima beans, peas, lentils, ratania, zarzaparrilla, vegetable ivory, archil, cochineal, cascarilla, cundurango, quillai, caucho, Peruvian balsam, tamarinds, mahogany, and different other products.

The fruits grown there are bananas, dates, cocoanuts, water-mellons, oranges, limes, figs, mangoes, pacaes, grapes, papayas, chirimoyas, pineapples, huanabanas, alligator-pears, guayabas, pomme-granates, plums, pepinos, prickly-pears, citrons, quinces, granadillas, lemons, etc.

There are also an infinite number of medicinal plants.

Cattle, goats, sheep, horses, mules, donkeys and pigs are raised there.

There are many gold, silver, iron, lead, copper, sulphur, and coal mines; lime and chalk quarries; asphalt, ochre and salt deposits.

There are more than 150 towns, big and little, in the Department.

The inhabitants are chiefly engaged in agriculture, and sow cotton principally, which grows in great abundance, and is of superior quality.

There are seventeen cotton gins working there, all the year round, the largest being at Sullana, Piura, Catacaos, and Querecotillo.

Trade is active, and, in years when the rainfall is abundant, may be estimated as follows:

*Exports.*

PAITA CUSTOM HOUSE.

| Per annum. | Value. |
|---|---|
| Cotton, more or less, 60,000 quintals.. S. | 1,200,000 |
| Straw hats, 26,205 kilos................ ,, | 369,092 |
| Cottonseed, 1,592,727 do.............. ,, | 32,961 |
| Tobacco, 309,341 do..................... ,, | 137,443 |
| Goatskins, 233,173 do .................. ,, | 257,811 |
| Carried forward............. | 1.997,307 |

|  |  |  |
|---|---|---|
| Brought forward........ | S. | 1.997,307 |
| Cattleskins, 54,590 do ............... | ,, | 26,323 |
| Coffee, 20,425 do......................... | ,, | 10,990 |
| Kindling wood, 1,262,000 do......... | ,, | 7,733 |
| Charcoal, 3,275,550 do................. | ,, | 50,560 |
| Wool, 4,172 do............................ | ,, | 2,495 |
| Cundurango, 2,746 do.................. | ,, | 23,570 |
| Algarrobo, 10,000 quintals............ | ,, | 15,000 |
| Salt, do....................................... | ,, | 150,000 |
| Fatted cattle, 4,000..................... | ,, | 136,000 |
| Asphaltum, and other articles ...... | ,, | 150,000 |
| Total............... | S. | 2,570,556 |

### TUMBEZ CUSTOM HOUSE.

|  |  |  |
|---|---|---|
| Tobacco, when the crop is good, 5,000 quintals............................ | S. | 125,000 |
| Whale oil..................................... | ,, | 100,000 |
| Kerosene from Zorritos................ | ,, | 186,000 |
| Mangrove bark............................ | ,, | 20,000 |
| Lumber........................................ | ,, | 10,000 |
| Hides, fruit and other articles......... | ,, | 20,000 |
| Dyer's Orchil............................... | ,, | 4,000 |
|  | S. | 465,000 |

### TALARA CUSTOM HOUSE.

|  |  |  |
|---|---|---|
| Kerosene and its products............. | S. | 300,000 |

### *Resumé.*

|  |  |  |
|---|---|---|
| Paita ........................................... | S. | 2,570,556 |
| Tumbez ....................................... | ,, | 465,000 |
| Talara ......................................... | ,, | 300,000 |
| Total exportation......... | S. | 3,335,556 |

The value of imports last year amounted to 1,099,809,50 as follows:

|  | | | |
|---|---|---|---|
| From | England | S. | 674,278.50 |
| ,, | United States | ,, | 195,195 — |
| ,, | Germany | ,, | 93,441 — |
| ,, | France | ,, | 70,675 — |
| ,, | Ecuador | ,, | 19,671 — |
| ,, | Chili | ,, | 17,558 — |
| ,, | Colombia | ,, | 7,919 — |
| ,, | Spain | ,, | 6,380 — |
| ,, | Italy | ,, | 6,019 — |
| ,, | St. Thomas | ,, | 5,205 — |
| ,, | Belgium | ,, | 3,468 — |

Total Importation......... S. 1,099,807.50

The duties collected at these custom houses amount approximately to S. 400,000.

With regard to roads, they are, generally speaking, good. Those on the coast are level, and, although those in the interior run through uneven country, they are easy to travel over.

Only two railroads have been built in the Department up till to-day. One, belonging to the Government, and let out to Don Federico Blume, for twenty five years, runs from the port of Paita, along 98 kilometres of level ground, to the city of Piura; and the other, 12 kilometres long, connects the latter place with Catacaos, and belongs to a joint-stock company.

The are two telegraph lines and one telephone line, with 150 miles of wire.

Provisions, throughout the Department, may be had in abundance, and very cheap, especially meat, which is of superior quality.

Laborers are found for all kinds of work, at the moderate wages of 40 cents to S.1 a day, excepting at sowing and reaping times, when it is a question of supply and demand.

The proposed irrigation schemes will reclaim 1.000,000 hectareas, the equivalent of 2.500,000 acres.

Petroleum in Peru has been freed from all taxes or duties for the next twenty five years, according to the Act of Congress of November 8th, 1890, which is as follows:

"The Congress of the Peruvian Republic,

Considering:

That it is necessary to promote the investment of native and foreign capital in the mining industry of the country,

Has given the following law:

Art. 1. The taxes which to-day burden mining property and products shall not be augmented for the term of twenty five years, counted from the date of this act, nor may other new ones be created during that period.

Art. 2. By the word "products" are meant, for the purposes of this law, gold, silver, copper, iron, tin, nickel, antimony, cinnabar, coal, sulphur and PETROLEUM, in their primitive condition, as well as worked.

Given etc."

According to the text of this very liberal law, which is an honor to the Legislature that passed it, no new duty can he created without the prestige of Peru being ruined, and without there being any means of preventing European capital from being frightened out of the country forever.

We will not close this chapter without transcribing the passage, in its entirety, of the work of the French writer, M. Simonin, which treats of the industrial development of the United States of North America.

"If Nature has done much for the United States, men have done the rest, assisting, with their character and institutions, the development of the immense natural wealth throughout the land.

"In North America, the individual is everywhere, and the Government nowhere. The initiative of the citizen is never restricted in the least. The hand of the Government, when it is felt, is only to help individual labor; to guide it with valuable and important reports, statistics and publications, carefully compiled; but never to molest it with a series of tedious, minute and useless formalities, the tradition of which the latin race piously keeps up.

"In North America, nothing lies locked up in the national archives: everything sees the light of day, promptly and at the desired moment. Industry is free, and monopoly unknown.

"It is for this that individual initiative has done such great things in that privileged land, and given the petroleum industry that healthy impulse, the results of which we are benefitting by".

What a good example to imitate!

*TABLE showing the owners of Petroleum Claims in the three petroleum regions of the Department of Piura.*

## REGION OF THE CENTRE.

(PAYTA).

Oscar Heeren, Enrique de la Riva-Agüero, Enrique Prevost, Cárlos

| | | |
|---|---|---|
| Leguía, Antero Aspíllaga, Isaac Alzamora, Lorenzo Barbera...... | 200 | CLAIMS. |
| Francisco Mario de Albertis, Guillermo Ferreyros, Cárlos Ferreyros, Manuel Zevallos, Manuel Moreno y Maiz, José Buccelli, Juan F. Valega, Lorenzo Barbera, Cárlos F. de Albertis, partners of the Peruvian Petroleum Company.... | 250 | " |
| Rafael Rojas Quezada, Lorenzo Barbera & Co.,........................ | 200 | " |
| Juan Bautista Serra, Lorenzo Barbera & Co.,.......................... | 250 | " |
| Federico Moreno & Co............... | 80 | " |
| London Pacific Petroleum Co....... | 10 | " |
| Mulloy & Co................................. | 9 | " |
| Arellano & Co............................. | 14 | " |

## REGION OF THE SOUTH.

### (SECHURA).

| | | |
|---|---|---|
| Blume, Dávila & Co., Cabieses and others ........................................ | 300 | " |
| Moreno & Co................................. | 270 | " |

## REGION OF THE NORTH.

### (TUMBEZ).

| | | |
|---|---|---|
| F. G. Piaggio.... ....................... | 54 | " |
| Lama & Co.................................. | 300 | " |
| Mendizabal Hermanos....... .. ...... | 112 | " |
| Moreno & Co............................. | 80 | " |
| Harris & Co................................ | 16 | " |
| Miranda y Lores................ ........ | 10 | " |

# PART THE SECOND.

## CHAPTER I.

*Industrial Development of Petroleum in the United States of North America.*

A knowledge of the progress of an industry in the country where it has reached its zenith, is a safe guide for the initiators of the same industry in another country.

It is for this that we believe it to be a matter of lively interest, and an opportune subject at the moment, to publish an account of the struggles and hardships of the pioneers in the petroleum fields, when the beds were discovered, and of its immediate industrial application, which is the most brilliant title to glory of those who, penetrating a totally unknown region, reaped every possible advantage from it, in a short time, due solely to their courage and daring.

The following is the interesting history of the discovery of petroleum in the United States, taken from different newspapers and books which we have before us.

The first petroleum company organised in the United States, and called "Columbia," was authorised by the Legislature of Pennsylvania on May 31st. 1861, with a capital of $. 25,000 divided in 1000 shares of $. 25 each.

They began operations at Story, on Oil Creek, seven miles from its source.

During the year 1861, the output of petroleum was 20,000 barrels; in 1862 it increased to 89,602 barrels.

Its first dividend was declared on July 8th, 1863, a little less than two years from the date of its installation, and was 30 per cent on its capital stock. This was followed by another dividend of 25 per cent., on August 12th; another, at the same rate, on September 9th; and still another, on October 14th, of 50 per cent, making the total of the dividends paid within two and one half years of the Company's installation, 130 per cent on its subscribed capital.

In 1864, the output increased to 141,508 barrels.

During the first six months of that year, it declared four other dividends, amounting to 160 per cent on its capital. At that moment, the capital was increased to $. 2.500,000, and a dividend immediately declared on the increased capital of 5 per cent., and, before the end of the year, five dividends more, amounting to 35 per cent in all. From that time to the end of 1871, the output of petroleum was always about the same, the minimum being 110,655 barrels in 1867, and the maximum 142,034 barrels in 1871.

The total output of this company, during those ten years of its existence, was 1.715,972 barrels, and the total amount of its dividends S. 2.342,600 or 401 per cent on its capital, and nevertheless, after ten years of active work, only a very small portion of its land had been touched, and the part that has been worked will admit of even more wells being sunk in it in addition to those that already exist there; and, although the golden dreams of the shareholders have been more than realised, there is no reason to doubt that these dividends may be maintained or increased, at pleasure, in years to come.

The history which we have tried to sketch of this company, is of numbers and not of words; but from these numbers an idea may be formed of the vast wealth which still remains hidden in the cavities of the rocks, and which only requires the hand of science, and the direction of prudent administrators, to place it within the reach of man.

A part of the land of this company was sold for $. 8.000,000 sols.

*The Oil Fields.*—The most important petroleum beds in the United States lie in the western part of the State of Pennsylvania, on the northwestern descent of the Alleghany mountains and lake Erie, and extend northward to and penetrate the State of New York. This region, which is watered by the Alleghany river and its numerous tributaries, is rough, mountainous and covered with thick woods; the subsoil is composed, in great part, of porous sandstone, covered with coarse sand, the layers running in a southerly direction.

The whole extent of the oil fields is not over a hundred and fifty miles long, with a variable width of from one to twenty. Oil Creek, a tributary of the Alleghany to the east, and French Creek to the west, were for a long time the limits of the territory worked.

The petroleum industry was born at Oil Creek, where the first wells were sunk, and where, for many years, oil men concentrated their efforts.

Besides the Pennsylvania fields, others are worked

in West Virginia, Ohio, New York, Kentucky, Kansas, Colorado, Michigan, Indiana, Illinois, Utah and California. These districts being of little importance, they figure very low in the total production of the United States.

It is impossible to fix the date on which petroleum was first discovered in Pennsylvania, for there are reasons to suppose that the mineral oil was employed by a race living there before the indians of our epoch. In fact, signs of wells have been discovered in the ruins attributed to the Mound Builders, a mysterious people, known only by the remains which they have left scattered over the immense country watered by the Mississippi, Missouri and Ohio rivers.

It was in the course of the year 1845 that Mr. Peterson, the proprietor of some land on which there were a few salt water springs, while digging a well with the object of finding fresh water, found, with astonishment, instead of water, a blackish, oily liquid.

Analised by the director of the Hope cotton factory of Pittsburgh, he discovered that through a chemical process, and mixed with spermaceti, the new substance constituted one of the best lubricants for machinery. The commercial value of the mixture was barely 70 cents per gallon, while the spermaceti cost $.1.30 per gallon. The economy was so manifest that from that moment ten barrels per week were contracted for consumption in the factory.

For ten years, the proprietors of the Hope cotton factory employed this greasy petroleum, without anybody suspecting its composition. This was the first use to which petroleum was put in Pennsylvania.

Nevertheless, the illuminating properties of this oil were known in other parts of North America, for, in a letter addressed by Dr. Hildreth to the *Journal of Sciences*, in 1826, he asserted the fact that petroleum might be employed for illuminating purposes, and that, later on, it would be of great utility in that direction, in the future cities of America.

In 1858, the city of Pittsburgh was lighted with refined petroleum, the same process being employed as was used in the distillation of bituminous schistes, by James Young, of Nova Scotia. This oil was known by the name of Carbon Oil, and the consumption was so great that, in a short time, it absorbed the output of the well christened Tarentun.

The numerous attempts to refine oil, as a necessary consequence, attracted the attention of capitalists and business men, who began to look around for the new substance.

Investigations, which were at first carried out all over the state, were soon concentrated around Oil Creek, where petroleum overflowed the surface of the ground and was collected by a New York company called the Pennsylvania Rock Oil Company.

In 1858, the directors of this company secured one hundred acres of land near Venango, a mile and a half below the well which had been worked for centuries by the indians.

Col. Drake, after having dedicated many years to the study of this substance, and suffering all kinds of reverses and deceptions, put himself at the head of the new works, with the well-founded hope of finding, at no great depth, the supply wished for.

Work began in June 1859; and in August of that year the drill pierced the bed at a depth of 182 feet. A pump which had been put in place brought the oil to the surface, and from that moment "Drake" well yielded 25 barrels every 24 hours.

From that memorable date, the new industry was definitely established.

The supply of the raw material being thus assured, the producers directed their attention to perfecting the system of refining it, and obtained, in a short time, an illuminating oil, economical, and of easy application.

Once the definite result was arrived at, the enthusiasm of the owners of the land around Drake well increased, and they began to dig wells for their own account, or ceded permission to do so in consideration of heavy premiums. Everyone went to work, and boring aparatus were put up daily by the dozen.

Some of the new wells were productive, but others dried up after giving a small quantity of oil, and were abandoned by their owners, who despaired of the result.

Until the discovery of *spouting* and *flowing* wells, which completely changed the aspect of the new industry, the demand for and the use of petroleum were limited; but from the date of the discovery of those wells which gave oil without any necessity of pumps or other expense of any kind, and in extraordinary abundance, the demand increased to such a point, that the oil was sold at 10 cents per barrel.

The greater number of the wells which required pumps to take out the oil were abandoned, and the owners, believing themselves ruined, were in despair; but these wells afterwards gave the most brilliant results.

Those who had been poor laborers, in those districts, became opulent millionaires through this unexpected industry, which increased the price of an acre of land from ten dollars to ten thousand dollars —a sum which the oil men cheerfully paid.

This sudden increase in the value of property was not the only consequence of the discovery of petroleum, for, as we shall see further on, colossal fortunes were improvised within a few months, while, at the same time, enormous losses perturbed the business and the lives of many Americans.

## CHAPTER II.

The first flowing well at Oil Creek was *Funk Well*, and it was sunk in the farm called *El hinney*.

Mr. Funk was a poor man when he began to bore, and it was amidst countless difficulties that in the month of June 1861 he arrived at the *stratum*. To the surprise of all present, the petroleum spontaneously rose to the surface, and the well from that moment yielded 240 barrels daily. Such an unexpected event as this broke up many combinations, and interested parties prognosticated that the well would not last long, and that no other would be found like it.

In the meantime, oil continued to flow from the well for fifteen months, at the end of which time Mr. Funk was a pretty rich man.

A few months before this well dried up, two more were bored, the *Philip* and the *Empire*, both in the same locality, and which yielded five thousand barrels daily.

The owners were thunderstruck and did not know what to do with such an enormous yield, the petroleum industry being then in its infancy. For want of receptacles they allowed the oil to flow from the wells until it converted the neighboring land into veritable lakes.

The opinions of the capitalists and of men of science on the petroleum question differed; many had doubts as to its future and of its value, and, above all, were resolved to withdraw their capital.

The five thousand barrels of oil, with which the flowing wells daily supplied the market, had brought down the price and left large quantities of oil undisposed of.

While all this was going on, the two new wells, *Sherman* and *Coquette*, were sunk, which increased the annual yield to 500,000 barrels.

After these wells, the *Agitator* was sunk, which it was necessary to pump for half an hour to make it give out oil for the succeding half hour. *Sunday* well yielded oil only on Sundays, when its neighbor the Agitator was not working.

In the meantime, the aspect of the market had changed completely. Consumption increased rapidly, and, while 500,000 barrels were sold in 1861, in the following year, 1862, 3.056,690 barrels barely supplied the demand.

New wells were bored in different places, especially in Pleasantville and Pithole, which, together with those at Tidioute, brought up the supply in a short time, to 3.887,700 barrels.

The new wells, *Venango*, *Buiter* and *Clarion*, in 1874, produced 10.809,852 barrels. After this period of rapid increase, came a notable falling off.

In 1875, Bradford began to feel the results obtained from its flowing wells; and, in 1880, in the total production of 26.000,000 barrels, Bradford figured with 20.000,000.

The Alleghany wells, opened in 1881, followed those of Bradford in importance, and with this reinforcement, the yield, in 1882, reached 81.398,750 barrels, or over *five thousand million* litres!

The circunstances attending the discovery of these beds, and the consequences to the petroleum industry, make it worth while to give a few details about the sinking of the first well.

Cherry Grove, in Warren county, was, until the month of April 1882, almost a deserted district, surrounded by thick woods, and inhabited by hardly ten or twelve people. The small town of Clarendon, a station on the Erie railway, is situated ten miles away from this place, and four miles away there is a small oil refinery.

For a long time before, oil prospectors asserted the existence of large quantities of petroleum southeast of Clarendon; but, notwithstanding such predictions, no speculator cared to undertake work there.

In April 1882, four men, more adventurerous than their brothers in the profession, established themselves at Cherry Grove, secured a few acres of land for cutting kindling wood, and secretly began the work of sounding. Once it was known that these wells were bored, there was a presentment as to the result.

Stockbrokers, speculators and, in general, everyone interested in the petroleum industry, sent their spies to find out how the work was getting on; while the adventurers, in their turn, took all possible precautions to cover their tracks.

In spite of everything, however, there was a courageous young fellow, who, at the risk of his life, and quicker than they, succeded in hiding himself for ten hours, under the platform of the well, and gave his employers the information that the like of well N°. 646 had never been seen before. This number, by which the well was designated, was the number of that particular lot of ground in the rural registers.

The fortunate owners of the bed made the results known, and it was soon discovered that the well yielded 4,000 barrels of oil per day. This unexpected result produced a serious disturbance in the market, and the price of kerosene, which was then quoted at $ 1, fell to 45 cents.

From May 17th, 1882, the day on which petroleum was first taken from well 646, until the end of June of the same year, 320 wells were bored, which all gave a similar yield.

This prosperity, however, was not of long duration, and in a short time the supply of oil became exhausted. By the end of October of the same year, the yield was almost nil; the wells were abandoned, the woods were deserted, and, with the disappearance of the wealth of the district, there disappeared also the fortunes of thousands of individuals who had founded the most brilliant hopes on Cherry Grove.

Since that time, new beds are discovered daily, and new wells are bored of more or less importance.

Thus the *Christi* well, which has not its equal in size in the world, poured forth 6,000 barrels of oil in five hours. The 39 wells of Oil City give 4,200 barrels per day, and there are others which yield as much as 170 barrels per hour.

At present, 25,000 wells are being worked in Pennsylvania, the output of which is estimated at 40.000,000 barrels per year, or the equivalent of 6.400,000,000 litres.

The number of wells abandoned on account of the supply being exhausted is incalculable.

*Prospecting for Oil, and Boring Wells.*

Excepting in rare instances, the presence of petroleum is not manifested, on the surface, by any particular sign. Filtrations through crevices in the rocks, and escapes of gas above ground are not frequent in Pennsylvania, and chance is, consequnetly, almost always the rule in its discovery.

The depth at which petroleum is found is very variable and uncertain. It depends, in general, on the exterior configuration of the ground, and also on the dip of the stratum. Oil is found at a depth of from 100 to 2,000 feet.

When prospecting first began, it was thought that the vein followed the course of the depressions between the rocks that formed the gulleys, and wells were opened on the sides of the latter. Later on, this idea was abandoned, and the centre of deep gulleys was preferred. Afterwards, with more perfect machinery, wells were sunk from the heights, until at last the configuration of the ground was not taken into account at all. What are, however, five hundred feet, more or less, with such perfect apparatus as is now employed?

## CHAPTER III.

### *Prospecting for Oil.*

Prospecting for oil in Pennsylvania is done by so-called *wild-catters*, who carry on their business either for their own or somebody else's account. Like the gold prospectors who penetrated unexplored regions, examine the quartz and oil, and wash the sand on the river banks, wild-catters visit places where they believe oil is likely to exist, examine the ground, make soundings more or less deep, and, finally, when they think the right moment has arrived, sink wells to discover the so much sought-after oil.

Those active, intelligent and indefatigable men,

leading a life full of privations, hardships and constant adventure, rich to-day and poor to-morrow, bear up stoically under their changes of fortune.

Many of these people thought, at first, that an infallible indication of the presence of oil was, a certain characteristic odour rising from the grounds. These were called *oil smellers*. Others used magic rods and cabalistic words to determine its locality. Once placed, almost always, wtih precision, the wooden derrick, 80 feet high, is put up, and in this works the drill, within an iron tube, either by hand or steam power.

This drill, in order to reach the petroleum bed, has to pierce three successive strata of sandstone, according to the geological theory of the experts. The first, immediately after the layers of alluvial soil; the second, at a depth of from 100 to 300 feet, and in which the first indications of petroleum should be found. Sufficient oil is frequently met with in this stratum to avoid the necessity of going deeper; but, as a general rule, it is prefered to go down to the third stratum, where the oil is undoubtedly discovered in greater abundance.

During the last few years, a fourth stratum has been found in some places.

The work of boring is subject to many accidents, the most common, and, at the same time, serious of all being the breaking off the point of the drill, which if it sticks in the rock, makes it necessary to abandon the work.

When, as yankees say, oil is struck, a pump may be required to bring the oil to the surface, and the well is called a *pumping well;* or the oil may, of itself, rise to the surface, as in the case of a *flowing well;* or a column of oil may spout forth, as it sometimes does, 300 feet high, and then the well is called a *spouting well*.

If the well opened is a pumping well, the petroleum does not show at the surface, and it is gas only that escapes, with a peculiar sound, which willd catters have baptized *earth sighs*. A steam pump brings up the oil and deposits it in iron tanks.

If, on the contrary, a spouting well has been discovered, as soon the drill reaches the cavity containing the oil, the latter, impelled by the gas, rushes to the surface, and rises, in the form of a thick column, as we have already said, to a considerable height, in which case the mouth must be immediately stopped up, which is done with an apparatus invented for the purpose. The stream of oil frequently rushes forth with such violence that it throws, to a great distance, the drill as well as the derrick that supports it.

In the case of simply a flowing well, i. e., when the gas merely forces the oil to the mouth of the well, it is easy to screw in the stopper, which is worked with a key. This is the most economical class of all the wells known.

The natural gas that issues from the wells is used for illuminating purposes, and is employed as a motor for all the machinery of the establishment, it being collected in the generators after passing through a metal plate, perforated like a sieve, and of peculiar form.

The oil is deposited in enormous tanks, like the gasometres of large cities, and led from there, through pipes, to be distilled.

These tanks must of necessity be perfectly airtight, as the gas given off by petroleum is highly inflamable, and the slightest piece of carelessness may bring about an explosion.

The wells themselves sometimes take fire, especially the flowing wells, in which case, it is of no use to pour water into them, but rather stuff the mouths with earth.

One of these wells caught fire in 1882, at Cherry Grove, the fire lasting for many days and presenting the glorious spectacle of a column of flame, a hundred feet high, illuminating the extensive woods around it with a lurid light.

The cost of boring a well is estimated at $ 4,000, provided it is not over 500 feet deep.

In the work of transporting petroleum to the refineries and ports of embarkation, wonders have been accomplished; and, regarding the *pipe lines*, through which thousands of pumps daily force the oil, and which are extended all over the petroleum fields, one must admire the progress of this important industry which has improvised so many millionaires.

For the establishment of refineries, on a large scale, sites have been chosen, not in the centre of production, but rather at the ports of embarkation. The question of transportation has always been one of the most important in the petroleum industry and to-day Pennsylvania can show 4,000 miles of underground piping which connect and receive oil from the 25,000 wells that are being worked there.

There are five hundred iron tanks, with a capacity of 35,000 barrels each, with a thousand miles of telegraph line, communicating between them and the principal centres of production and refineries, which latter, in their turn, deposit daily in the tanks 200,000 barrels of oil, which is taken to the ports of embarkation in 2,500 tank-cars, hermetically sealed, all belonging to the United Pipe Lines Company, which, united, and forming a single body with the Standard Oil Trust Company, have monopolised the kerosene trade of the world.

The primitive capital of this company was 3,500,000 dollars.

Its mysterious organization does not allow of its details being publicly known, nor even how many shareholders belong to it. The administration of this gigantic business, directed by a group of bold speculators, is divided into large branches, or sections, each one representing the fourth part of the globe, which attend to the necessities of consumption, and carefully watch the market.

The founders of this cosmopolitan company have retired from active participation in the work, after making fabulous profits, and number amongst them Lokefeller and Andreau, both veritable money kings on account of their millions.

Up till to-day, North Americans wield the sceptre of this industry.

They have monopolised, and hold in their hands, all the markets of the world.

## CHAPTER IV.

*Russian Petroleum Fields and their Industrial Development. — Transcaspian Region. — Bakhu. — Caucasia.*

A French writer, treating of the study of pettroleum in Russia, expresses himself thus:

Towards the northwest, and following the chain of mountains, the first peaks of which begin to rise on the border line of Afghanistan, not far from the Caspian sea, are the Balkans, through which there flowed, in another epoch, the Oxus, and which is to-day the plan of the Trans-Caspian railway.

All this region, which extends as far as the sea and ends at the peninsula of Kradsnowodsk, is so immensely rich in petroleum that Russians have given it the name of *Black California*:

When the Russian engineers were making the survey there, in 1880, for the construction of a railway, on sinking wells for water, they discovered innumerable petroleum beds, in the place called Naphtha Mountain.

The layer of petroleum which covers the mountain measures no less than 10,000 hectares in extent, and could easily yield a million tons of petroleum per annum, or sufficient to illuminate all Russia, lubricate all her machinery, and move all her locomotives, merchant vessels and men-of-war.

This region, which, so far as petroleum is concerned, is one of the best in Russia, has been valued by the well-known Russian engineer General Bohrberg, and Prince Yeristoff, at £35 000,000; so that if this sum represents the value of only ten thousand hectares of petroleum fields, the million hectareas in Peru, in the same proportion, represent the enormous sum of £3.500,000,000.

Only a single well has been sunk in the place mentioned, which yields invariably 1000 litres per day, enough to satisfy the necessities of the inhabitants.

The Russians, after carefully studying this territory, look upon it as a reserve which they will work when their roads advance in Asia, when they will be able to make Merie a great petroleum centre, where the people of Central Asia, Afghanistan and Khorassane will be supplied.

On the shore opposite the island of Kradsnowodsk, there is a gigantic promontory, very high, covered with woods, and stretching out a long way into the sea. It is the peninsula of Aspheron, the holy land of the desciples of Zoroastro, the birth-place of the religion of fire-worshippers.

Looking from a distance at this tremendous cliff breasting the waves of the Caspian sea, when its wells spout forth columns of petroleum two or three hundred feet high, it resembles a gigantic whale, spouting streams of spermaceti from its wide nostrils.

At the lower end of a bay, formed and protected by two peninsulas of the island, lies Bakhu, the capital of the petroleum region of Caucasia, the richest and most important petroleum centre in the world.

Bakhu, with its white houses rising in rows one behind the other like an amphitheatre, at the foot of the green hills, its elegant cupolas, numerous bathing houses, splendid gardens, high observatories, narrow and winding streets, solid walls with the Persian arms, (*the Lion of Persia*)—Bakhu, the oriental city, will soon cease to exist.

Solid and elegant buildings have in great part replaced the old ones Wide and straight streets, paved with asphalt, to-day replace the narrow, dark and muddy streets of other times. And an immense embankment has been built in front of the town along the sea shore.

In a few years Bakhu will be a new city, with all the luxuries of modern civilization, and the comforts of the cities of Europe.

In 1870 it had hardly 12,191 inhabitants; and ten years later, in 1880, the number was estimated at 50,000.

This immense progress—this rapid transformation—is due exclusively to the petroleum fields, which are worked there on a vast scale.

Although these fields have been worked for many years past, the industry, hampered by the action of the Russian government, did not arrive at that stage of development, which it has reached to day, until after the monopoly to which it was subject was abolished.

The petroleum region of Bakhu, of which the Aspheron peninsula is the centre, lies between the Samur river and the Caspian sea, its extension being 200,000 square miles.

Petroleum was found in this region, when the first wells were sunk, at a depth of 259, 560, 286 and 350 feet. These wells yielded little oil; but another, barely 70 feet deep, had given oil in abundance for many years before. Under the impression that petroleum would be found at the same level, another well was bored near the one mentioned, and oil was only discovered at a depth of 420 feet.

The theory of one common bed or level vanishes and is refuted in face of the existence of flowing and spouting wells, which, notwithstanding their abundant yield, do not diminish in the least the product of the neighboring wells. Thus, while the *Droojba* well, to which we will refer later on, spouted out a column 300 feet high, as if it were emtying the inside of the earth, the wells around it continued to yield their ordinary quantity of oil.

The wells of Bakhu are found in two distinct table lands, with a distance of eight kilometres between them, crossed by a trail.

The number of wells that are being worked to-day is four hundred; but the quantity of oil they yield is enormous and much more than that given by the 25,000 wells in Pennsylvania. Amongst these wells there are several natural wells, some of them constant and others intermittent.

Amongst the spouters is the famous Droojba, which, as we have said, when it was bored, it spouted forth a column of oil, with a roaring noise, 300 feet high, higher than the famous Geiser in Iceland, inundating the whole country around.

In North America, this great well would have made its owner a millionaire, bringing in, as it did, £ 11,000 daily; but in Bakhu it was the ruin of the man who bored it, for he lost his whole fortune in compensating the damages caused to the farms in the vicinity.

For a long time, the Droojba yielded 2,000,000 gallons daily, which quantity was afterwards reduced to 250,000 gallons. It is stimated that its total yield, until it gave out, was 500,000 tons of oil.

The wells in Bakhu are at short distances from each other, there being places, within the space of a few square yards, where four wells have been opened, and which always yield oil, notwithstanding the fact that they are at depths varying between 250 and 560 feet, and amongst them one which, although only 70 feet deep, has yielded oil for several generations past. This proves that the deposits of petroleum are independent, of each other, and that it has yet to be discovered how deep they may be found.

One of the wells belonging to Messrs. Nobel Brothers spouted out 112,000 tons into the air before the stopper could be ajusted, an operation that lasted a month. Number 9, the name of another well belonging to the same company, yielded 8.000,000 gallons in twenty days, of which 1.600,000 gallons were sold, 1.200,000 gallons were stored in a depression of the ground, and 5.200,000 gallons were lost and the surrounding country inundated with the oil. The stopper was finally put on, and it proved to be one of the most productive wells.

On several occasions the yield was so abundant that the owners, for want of consumers, had to open ditches and run the oil into the sea.

M. Berthelot, treating of his theory on the mineral origin of Russian petroleum, says he has discovered a way to calculate the quantity of petroleum in the porous sandstone, which may contain oil lakes of 30, 50 or 100 feet in thickness, without any large caves or defects intervening.

An acre of ground has 43,560 square feet. A square mile, 27.878,400 square feet, or 4.074.489,000 square inches. A barrel of petroleum of 42 gallons has 9,702 cubic inches capacity, or 5 6/10 cubes.

Let us now see what would be the production of an acre in proportion.

If the bed of oil be one inch thick, it should give 646 barrels; if it be two inches thick, 1293; if three inches thick, 1939; and if it be 7¾ inches thick, 4997 barrels.

A bed of oil one square mile in extent, if it be one inch thick, should infallibly yield 414,779 barrels; if it be two inches thick, 829,559 barrels; if three inches thick, 1.244,338; and if 7¾ inches, it would contain 3.198,515 barrels of 42 gallons, which represent 137.337,630 gallons.

Experiments made by M. Berthelot show that the oil containing capacity of sandstone is equal to one eighth of its volume. Now then, taking each vertical foot of the thickness of the sandstone to be equal to a petroleum bed one inch thick, and taking into account the fact that in Pennsylvania the beds are from 30 to 50 feet deep, the result is that, accepting merely 15 feet as the average depth, every acre of petroleum ground is capable of yielding 15,000 barrels of oil, and each mile 9.600,000 barrels.

Bakhu has 200 petroleum refineries, and seven railways for the service.

Almost all the wells of this region are flowing

wells, so that great precautions have to be taken to prevent the oil from being lost.

The quantity of kerosene, apart from crude petroleum, produced by the 200 refineries at Bakhu is as follows.

| Years. | Tons. |
|---|---|
| 1872 | 16,400 |
| 1873 | 24,500 |
| 1874 | 23,600 |
| 1875 | 32,600 |
| 1876 | 52,100 |
| 1877 | 72,600 |
| 1878 | 97,500 |
| 1879 | 110,000 |
| 1880 | 150,000 |
| 1881 | 183,000 |
| 1882 | 202,000 |
| 1883 | 206,000 |
| Total | 1.180,300 tons |

in twelve years.

The production at the same place from 1883 to 1885 may be classified thus:

| 1885 | 1884 | 1883 | |
|---|---|---|---|
| 137.000,000 | 109.000,000 | 72.000,000 | gals. kerosene |
| 19.000,000 | 10.000,000 | 10.000,000 | crude oil |
| 170.000,000 | 142.800,000 | 87.000,000 | residue |
| 7.950,000 | 7.200,000 | 5.000,000 | lubric. oil |
| 140,800 | 380,100 | 240,000 | benzine |
| 334.890,000 | 269.380,100 | 174.240,000 | |

From the year 1886 up till the present time, the production has doubled.

## CHAPTER V.

*Petroleum Deposits. Refineries. Exportation of Oil.*

To the left of Bakhu, and facing the sea, there is the improvised city of Tcharny Gorod, the *Black City*, as it is called, from the smoke that pours from the chimneys of its two hundred factories, forming a peculiar atmosphere, with a nauseous odor.

Everything is sombre in the Black City; the sky, the houses, the ground (which, mixed with petroleum, forms a slippery mud) and even the skins of the inhabitants, naturally white, have turned black through contact with the atmosphere of oil and smoke.

In the centre of a sterile and barren region, where there is not the slightest trace of vegetation for many kilometres around, a number of low structures are seen, surrounded by wide and blackened walls. It is there where the immense tanks which hold the oil are to be found, and which are spread over the ground as far as the sea, to where the vessels are, which load petroleum for Tsaritzen, a port as dirty and greasy as the Black City, and where the streets, houses, furniture, docks, ships and the very sea are covered with a thick coating of oil.

The mode of refining oil is the same everwhere. Taken from the wells, and put into retorts, it is submitted to a constantly increasing temperature, through which it undergoes the different transformations we have referred to before.

The petroleum of Bakhu, in distillation, does not

yield such a large quantity of kerosene or illuminating oil as North American petroleum does. From the latter 70 to 75 per cent of kerosene is extracted, while the former hardly gives 27 to 30 per cent. Peruvian petroleum gives more than 60 per cent of kerosene.

This inferiority, although so great, is compensated by the extraordinary abundance of the raw product, its low price and the numerous applications given in Russia even to the residue.

All the petroleum produced in Bakhu is expended in the form of kerosene, and is consumed throughout the vast Russian territory, without any necessity for the American product.

Four large commercial centres divide the output: Moscow, St. Petersburgh, Varsovia and Saratoff, where reservoirs have been built containing each 82.000,000 litres of refined oil.

From Bakhu to the island of Taman stretch the vast petroleum beds of Caucasia, covering an extent of 2,400 kilometres long by 16 wide.

The beds discovered here so far invariably follow a southeast to northwesterly direction, in a straight line and parallel to those in the south of Caucasia. The greater part of the beds are being worked: Ter, with 120 wells, gives an annual yield of 400 tons of oil; while Tifles, with 28 wells, gives 2000 tons. At Kunado, 40 wells give, on an average, 4000 tons.

On the island of Taman, all over the lower part of Kouban, petroleum beds are numerous, and there is much analogy between them and those of Pennsylvania, the oil being found in strata of stone, in small cavities, instead of as at Bakhu, in immense sheets or natural reservoirs of great capacity.

From the foregoing it will be seen that there are petroleum beds, although irregularly distributed, under the whole length of the Caucasian hills. It is not abundant in the centre, while the extremes are incalculably rich. Throughout the length of the beds, the petroleum is found at different levels, varying from 200 metres below the level of the sea, to 300 above it.

The actual value in Russia of a square metre of petroleum ground is one pound sterling; so that our petroleum claims in Peru, which measure 40,000 square metres each, would be worth, in the same proportion, £40,000 per claim.

## CHAPTER VI.

### *Transportation of Kerosene.*

Nearly all the petroleum refined at Bakhu is disposed of, in the form of kerosene, throughout the Russian empire. To carry out this complicated operation, there are only two routes: that of the Caspian sea, and that of the river Volga, the latter being 2400 miles long, and over which specially constructed steamers run as far as Tsaritzin; while the former goes via the Caspian sea to the ports of the Black sea.

When work first began, transportation was effected in wooden barrels, at enormous expense and with endless difficulties, owing to the inuumerable series of trans-shipments, and the high rates of freight over the railways.

Apart from all this, the refining companies constantly suffered heavy losses from the very nature of the barrels, their weakness and bad make.

The petroleum industry was on the verge of ruin, when Messrs. Nobel Brothers, the largest refiners, proposed to the Russian company that has the privilege of navigation on the Caspian sea to construct a certain number of tank steamers, on the most advantageous terms to that company; but as they refused the proposal made by Nobel, the latter built the first vessel of this kind on his own account. The result was splendid, and a short time afterwards those wealthy business men had at their disposal twelve big steamers, of the same class, each capable of carrying 750 tons of kerosene.

Induced by this example, and above all by the results, the other refiners in Bakhu commenced to build steamers, and to-day this fleet in the Caspian sea exceeds one hundred tank steamers. The operation of loading and unloading is simple in the extreme.

The steamers fill their tanks by means of long hose from the petroleum reservoirs at the Black City; when loaded they cross the Caspian sea to the mouth of the Volga river; but, as the bar of this river is dangerous to vessels of deep draught, the oil has to be trans-shipped to smaller vessels, with the aid of hose and pumps. These cistern steamers easily run down the river as far as Tzaritzin, 400 miles from its mouth, to the first railway station. The voyage is generally made in two days, and on the arrival of the cisterns the oil is pumped into the big reservoirs on land, from which it is sent to all the markets in Russia.

The intrepid Messrs. Nobel, Bros., having obtained this advantage, it was necessary to secure another, which consisted in transporting the petroleum over the railways in the same way as it was done by the steamers, that is to say in tankcars, an improvement which Messrs. Nobel finally brought about, building fifteen hundred of these cars, which run, without interruption, over the immense network of Russian railways; the loading and trans-shipment being effected at night, the scene being illuminated with the electric light.

In order to facilitate serving out the oil, large reservoirs have been built on the banks of the Volga, as well as at distant points, the former having a capacity of 22.717,250, and the latter 162.755,510 litres.

The four great kerosene centres are: Moscow, St. Petersburgh, Varsovia and Saratoff. Between these, the Black Sea and Germany on the one side, and the river Volga on the other, there are 21 stations of less importance. During the summer, sixty trains, which start from the Volga, supply the 36 stations distributed over Russia in Europe, Poland and Finland.

The Nobel Company does not sell oil by barrels, but by tank-carloads, delivered at the station desired by the purchaser, where it is paid for cash down, this big company not selling a drop of oil on credit.

In the central office at St. Petersburgh, telegraphic despatches are received from the conductors of the trains, the positions of which are indicated with little red flags on the great map of Russia; so that during the whole year, the positions of the sixty oil trains running over the lines can be known in an instant— by a *coup d'œuil* as it were—by the Manager of the Company.

The tank-cars generally contain ten tons, twenty five cars making a train. Each wagon is filled or emptied in three minutes and a half, and the whole train in an hour.

## CHAPTER VII.

*Petroleum Worship.*

The history of petroleum in Russia is extremely curious, for, in remote ages, it was the firmly established religion of various sects.

There can be no doubt but that Fire-worship, of which Zoroastro, according to some, was the founder, and, according to others, the restorer, sprang into existence at Bakhu.

The aspect of the island of Aspheron, crowned with fire, illuminating at night the waters of the Caspian sea and the bare country around it, was more than sufficient to excite the imagination of the superstitious orientals.

We must not therefore be surprised if they took for a divine manifestation a physical phenomenon, unknown before, and which, at that epoch, nobody was able to explain.

In the time of Cyrus, the followers of Zoroastro were already very numerous, and every year they went to Bakhu, in great caravans, to prostrate themselves in the temple, on the altars of which burned the eternal fire. These long pilgrimages lasted until the day Heraclites invaded Persia, and, giving himself out as a christian king, extinguished the sacred fire, destroyed the temple, and dispersed its priests. This is recorded by Gibbon in his work "Decline and Fall of the Roman Empire". This religion lasted 2,000 years.

Twelve years later Persia was conquered by the Arabs. The conquerors obliged the adorers of the eternal fire to adopt Islamism: many submitted, but the greater part emigrated to India, in order not to accept the new religion, and formed the sect called Parsis.

Until the beginning of the twelfth century Bakhu was visited as the birthplace of the sacred element.

The phenomenon that gave rise to the formation of the religious sect in question was no other than the column of petroleum which burned constantly, while the ground around it, saturated with the oil, formed a perfect sea of fire.

In the thirteenth century, Marco Polo wrote upon the medicinal properties of this substance, which was eagerly collected by all the inhabitants of the neighboring country.

When Peter the Great, in search of conquest, proposed to make the Caspian sea a Russian lake, he directed all his efforts against Bakhu, the wealth of which his eagle eye had been able to appreciate. After its annexation to the Empire in 1723, he gave special orders to have petroleum extracted and sent to Russia via the Volga.

An Englishman, Jonas Hanway, entrusted by his government with a special mission to the shores of the Caspian sea, visited Bakhu in 1754, and was the first to describe the richness of those petroleum beds, and the uses to which the oil was put by the natives, asserting, in his famous book, that petroleum was an excellent remedy against stone, affections of the chest, and pains in the head.

Until a few years ago, three priests officiated in the temple of the eternal fire, two of whom died of old age, and the third was murdered. A refinery was then established on the site of the temple, and it turns to profitable account the petroleum aud natural gas that escape from the wells.

Petroleum, as far as its uses are concerned, was perfectly well known in ancient days, and is as old as the world.

The asphalt lake, the Dead Sea, is nothing more than an eruption of petroleum, and some authors believe that that eruption caused the destruction by fire of ancient oriental cities.

It is the same substance as the Egyptians call *balsam of mummies*, which was taken from the Red Sea, near Suez, and which they employed to embalm their dead, covering them with a winding sheet soaked in oil. Even to-day, notwithstanding the forty centuries that have passed, the imperial mummies discovered in Egypt, preserve the pungent odor of petroleum.

The cement employed in the walls of Nineveh and Babylon is a composition of which petroleum forms a large proportion. It was taken then from the beds at Is, 180 kilometres from the last mentioned city, which beds also attracted the attention of Alexander and Trajanus. In the building of all the monuments erected by the latter, a cement was employed identical with that used in the walls of Nineveh, petroleum forming the greater part of it; and it is the same as that employed in the construction of the large cities on the banks of the Euphrates.

Herodotus speaks of the mineral oil of Zazin, the modern Zante of the Ionian sea.

Plutarch describes a sea of fire near Ecbatana; and Pliny, Diodorus, and Dioscoridus mention, in their works, the wells of Agrigante, from which oil was extracted that was burned in the lamps of Sicily, before the ruin of Pompey and Herculand.

Italy used crude petroleum for illuminating purposes for many centuries.

The mineral oil has been known, under different names, for centuries. It is found distributed all over the face of the globe. Its extraction is not always easy, as it is found at a great depth from the surface of the earth; but in all latitudes and in all places, when it can be got at, it is used to great advantage.

It is light, heat and fortune.

It has dethroned gold, and it is the most terrible enemy of coal. Coal is tired after its long pilgrimage on earth, during so many centuries, and must, in the end, yield to its rival.

## CHAPTER VIII.

### *Monopoly.*

Until a few years ago, in Russia, where the form of government, absorbing and despotic as it is, is so favorable to monopolies, the government had monopolised the petroleum industry. It was declared free in 1872, and since exonerated from all taxes or duties, present or future, the trade at that time amounting to 117,000 roubles ($. 87,000 gold) per annum.

In this respect, a well-known French writer, in treating of this subject, expresses himself thus:

"It is a fatal law, in everything and everywhere, that monopoly kills the industry it governs.

"Bakhu, the great petroleum centre of Russia, was not exempt from the consequences of that system.

"Although the suppression of the monopoly dates from the year 1872, a heavy duty burdened the product of the wells until 1874, and it was from the date of its abolition that the development of the great industry began, which is called upon to convert that far-off corner of the Caspian sea into one of the most powerful commercial centres of the world.

"A few figures will suffice to prove our assertion.

|  | Tons. |
|---|---|
| "In 1872, when the petroleum industry was declared free, the output per annum was | 24,800 |
| "When the duty was suppressed it reached | 242,000 |
| "And since 1879 it has risen to | 800,000 |

" The number of wells increased in the same proportion. In 1872 there were only two wells; while in 1879 there were 101, and in 1883 as many as 400 were opened.

"The complete liberty of the industry, and its exoneration from all duties or taxes, has assured the future of the petroleum industry in Caucasia forever."

Until 1883, in order to reach the heart of the Empire and approach the ports of Europe, the liquid combustible had to traverse more than *two thousand miles*, going up the Volga in special steamers; whereas to-day, with the termination of the Trans-Caspian railway, which was brought about by the liberty of the petroleum industry, 560 miles are economised, and the markets of the world opened to the products of the Russian Empire.

# PART THE THIRD.

## CHAPTER I.

*Scientific Aspect, Origen and Chemical Composition of Petroleum.*

Very little has been written on this important subject: a little manual in Germany; a good book containing a practical study of the industry, published in Philadelphia in 1887 by Benjamin J. Crew; and two which have been published in France, the better of them being in our opinion, that published in Paris, by M. Hué, in 1885.

From this book we take, entire, the technical part referring to petroleum, which is the subject of this chapter, and which we have translated literally from the French.

Petroleum, which the Romans called *bitumen*, owes its name to two latin words: *petra*, stone; *oleum*, oil—i. e. stone-oil. The abbreviation and union of both, in latin countries, has given the word *petroleum*.

It was already known to scientific men in the XV and XVI centuries by that name.

In North America, it is also called *petroleum*. In Caucasia it is known as *naphtha*: and in Asia it preserves its former name of bitumen.

Crude petroleum is a dark, greasy, viscious and opaque liquid; it has a strong smell, very pungent, and similar to heated bitumen.

It is composed of different hydro-carbonides, containing in solution parafine and asphalt.

Lighter than water, it distils without undergoing any change, but becomes limpid and colorless. It is in this state that it is handed over to commerce, as kerosene.

Chemical analysis reveals the presence of several substances in petroleum, their proportion varying sensibly, according to the spot it is taken from, and above all the depth of the bed. These substances are:

Light oils (illuminating),
Heavy oils (lubricating),
Parafine,
Asphalt, and
Residue (combustible).

At first sight it would appear that, knowing exactly the elements of which petroleum is composed, science should be able to formulate the correct theory of its origin, nature and formation; but, on the contrary, up till to-day, nothing certain can be affirmed, notwithstanding the serious investigations of geologists and chemists; and absolute obscurity prevails respecting this phenomenon, which has given rise to the most contradictory opinions.

According to many geologists, it is a liquid secreted by seams of coal, and forced above ground by the pressure of its gases. According to others, who believe the contrary, it is affirmed that coal is a product or transformation of petroleum.

In support of these theories, the partisans of both of them cite the indentity or similarity between the mineral oil and the oil obtained by the distillation of coal; but to this it is answered that in many regions where petroleum exists it is not accompanied by coal, and that the stratum in which oil is found in greatest abundance is below the coal-beds, while, in many places, it even takes a diametrically opposite direction. In the districts in Pennsylvania, where oil is found most abundant, it happens to be met with outside of the range of the coal deposits.

According to Reichenbach, petroleum is the turpentine of the pines of the corresponding geological epoch. M. Lesquerent believes it to be nothing more than coal in a liquid state, and, like the latter, a slow decomposition of vegetable matter, especially marine plants of cellular texture.

The gas produced by the fermentation of this vegetable matter remains enclosed with the oil in the cavities of the rocks, and the salt water, which is always found with the gas and the mineral oil, is nothing more than the remains of the sea water which covers the part where the *fucoides* abounded. Thus, when the drill pierces the stratum or rock where the petroleum is found, we see the oil mixed with gas and salt water issue forth like water from an artesian well.

The theory of the celebrated botanist is accepted by Bischof, and it is not far off from the opinion given by Daubrée, who expresses himself as follows:

"Although petroleum has been studied in different places, until to-day its origin has not been discovered with any degree of certainty. It is generally supposed that it is the result of the decomposition of marine plants and of animals that existed on the banks of the primitive seas. This hypothesis explains the presence of salt water and salt *gema*, the result of the sea water and organic remains having been shut up in the same cavities."

"A certain number of geologists, basing their opinions on the similarity frequently observed between the different strata of salt, sulphur and bitumen so often found in connection with the phenomena of dislocation, give petroleum a purely volcanic origin.

This theory has been accepted by Dufresnoy, Newbury and other scientific men.

Amongst all these authorities, at the time of the first operations in the petroleum fields of the United States, when everybody was giving an opinion as to the origin of mineral oil, there was not wanting a rather learned American who affirmed, with incredible effrontery, that petroleum was simply the urine of whales, deposited at the North Pole, and conducted to Pennsylvania through underground passages.

By way of throwing more light on the subject, we will cite the opinion of the Russian chemist Mendileef.

He presupposes the existence in the bosom of the earth of enormous masses of iron mixed with inorganic coal. Water rising to the surface comes into contact with the molten iron and is decomposed; the oxygen then combines with the iron, while the hydrogen, under the influence of heat and pressure, combines with the coal, and forms hydro-carbonate of petroleum.

Be it as it may, the fact is that specialists have entrenched themselves behind the theory that petroleum is the result of the decomposition or fermentation of vegetable and animal matter, through the agency of a phenomenon analogous to that to which the formation of coal is ascribed. There can be no doubt that as a consequence of this decomposition, certain volatile hydro-carbonides are formed, such as the gas of the swamps, giving petroleum those hybrid qualities which so greatly characterise it.

Another question of extraordinary importance, and which affects the petroleum industry in the highest degree, has presented itself at different times, and which is whether the formation of mineral oil, whatever may be its origin, can or can not be considered as a product of past centuries, or does it continue to be formed? In the latter case, the supply would be inexhaustable.

It is very difficult to decide this question. Nevertheless, there are reasons for believing that the existence of petroleum is not limited to the deposits discovered so far, and that Nature, which never paralizes her work of creation, is continually distilling further quantities of the precious liquid.

This opinion is confirmed by the fact that the wells which have become exhausted, when all their oil had been pumped out, have, after remaining idle a certain time, again produced oil in abundance.

Furthermore, we have the example of the wells of Birmania, Bakhu and Trinidad, which have yielded oil in abundance for many centuries, without the supply being diminished.

## CHAPTER II.

### *Where Petroleum is Found.*

As only negative results have been given by the investigations made, up to the present time, regarding the origin and formation of petroleum, we must, for the moment, look to the fields where it is found, for a rule which will allow us to discover, with any degree of certainty, the exact position of the deposits.

In many places, the existence of the oil is not manifested on the surface in any way, either by the presence of the liquid itself, or by the emission of gases.

The study of the fields that are being worked give results so diametrically opposed to each other, that it has been found necessary to abandon the attempt to find any indication of the presence of petroleum in the nature or geological formation of a given region.

This is just the reverse of what happens with many mineral substances, more especially with coal, which exists in certain well defined strata, while the liquid hydrogen is found in nearly all of them, without exception.

In Kentucky and Tennesse, petroleum is found in the lower silurean strata. The abundant supply in West Canada, belong to the lower devonian strata. The richest deposits in Western Pennsylvania belong to the upper devonian strata. Those of Virginia to the upper coal strata. In Connecticut and North Carolina, petroleum is found in the fissures of rocks.

The petroleum of Colorado and Utah is found amongst the lignites of cretaceous formations; and, finally, petroleum in California is found in formations belonging to the tertiary period.

Petroleum therefore runs through the geological scale, and is found in each one of its degrees. But we must take into account that in the richest petroleum fields of North America, the stratifications are all of the most ancient formation.

In Europe, Asia and Africa, it is just the reverse, Liquid mineral hydro-carbonides are found especially in tertiary formations. In Alsatia, Pechelbronn and Schwabviller, petroleum is found in tertiary formations. It is also found, under the same conditions, in Gabian and Herault. The most highly prized oil at Hannover is found in neocomic and jurasic strata. In Eastern Caucasia and coast of the Caspian sea, liquid hydro-carbonide is found in tertiary formations; and the same thing occurs in Galicia, Transylvania, Hungary, Birmania, Syria, Africa, and the coast of the Red Sea.

Throughout the Old World, where there is much more petroleum than in the New, the deposits of this oil are found in tertiary formations.

Nevertheless, whatever may be the nature of the soil which is bored through in order to reach the stratum, the oil is invariably found in sandstone.

Sometimes the strata are impregnated with the oil, and at others it is hidden in deep cavities, forming deposits or reservoirs, more or less extensive.

The thickness of the petroliferous strata, as well as the distance between them, are as irregular as the nature of sandstone itself. Some are dark, fine-grained and smooth, and others light yellow, coarse grained and very porous.

It is owing to these differences that we find such great variety in quality, density, color and consistency of the oil.

Petroleum, taken from shallow wells, rising of its own accord to the surface and issuing from sand or porous clay, is heavy, dark-coloured, and similar to coal tar; the oil from deep wells, like those of Pennsylvannia, some of which are *two thousand feet deep*, are much lighter in colour and in weight; it may be considered as having been distilled or purified by the heat of the earth, the heat of which is greater the deeper the strata. This is one of the great advantages of American petroleum, which contains from 70 to 75 per cent of light or illuminating oil.

To recapitulate: petroleum is not distributed over the sub-soil in the form of lakes, ponds or streams, but in the cavities and fissures in the rock, which it saturates and impregnates, mixed with gases and salt water. These substances are found in the order of their specific weights: the water lowest; on the top of it, the oil; and the gas compressed between it and sides of the rock.

These reservoirs, where the oil is found, are generally discovered in an oblique position. The drill may pierce, indifferently, the part containing the water, the oil, or the gas.

In the first case, the liquids, compressed through the pressure of the gases, are forced violently to the mouth of the well, when it is called either a flowing or spouting well.

If, on the other hand, the drill strikes the point where the gas is found, the latter issues in abundance, and it is necessary then to employ pumps, to bring the oil to the surface, and the well is called a pumping well.

The size of the cavities containing petroleum is altogether unknown; and it is also impossible to estimate their importance or duration.

A well, at a certain point, may produce a barrel of oil a day, while another, opened at a short distance from it, three or four thousand barrels every twenty-four hours. The yield of some may not be diminished for years, while others suddenly stop giving oil, or they are exhausted after being worked for a few weeks.

In North America it frequently happens that the wells on which the greatest hopes were founded, dried up within a short time, and caused the loss of fortunes. It is true, however, that in that country, the petroleum industry is a speculative one.

Considerable sums of money have been spent in sinking wells at the bottom of which not a drop of oil has been found. These the Americans call *dry holes*. At other times, the drill perforates the stratum at the very beginning, and the oil rushes out in torrents, making their fortunate owners rich in a few hours.

Millions of dollars have disappeared in fruitless explorations, and according to the exaggerated saying of oil men, the oil fields of Pennsylvania have swallowed up more money than they ever produced.

The following is a table showing the number of wells sunk which have and have not given oil:

| Wells opened. | With oil. | Without oil. |
|---|---|---|
| 1876 | 2.319 | 329 |
| 1877 | 4.056 | 657 |
| 1878 | 2.988 | 373 |
| 1879 | 2.798 | 141 |
| 1880 | 4.203 | 143 |
| 1881 | 3.848 | 167 |
| 1882 | 3.263 | 178 |
| 1883 | 3.949 | 263 |
| 1884 | 5.195 | 256 |
|  | 28.619 | 2.507 |

## CHAPTER III.

### *Refinement.*

To refine petroleum means to separate from it the volatile and inflammable substances which it contains: to separate the light from the heavy oils and parafine is to make of this viscious, dark, greasy, and extremely inflammable and dangerous substance, a transparent, inoffensive liquid, appropriate for burning in lamps, and giving an intense and brilliant light.

The separation of the different substances which it contains is effected by a series of special distillations, which give the different products that daily receive new industrial applications.

Petroleum on leaving the wells in deposited in large retorts and is progressively submitted to a temperature of 400°.

In order to avoid the ignition of the gases given forth during the operation, instead of heating it by fire direct, steam is employed.

By the action of the heat, raised from 15° to 180° Centigrade, the light or volatile parts are given off in the form of gas, which passes to the condensers and is condensed into different liquid hydro-carbonides, known to commerce as naphthas or essences of petroleum. Their density varies from 0.600 to 0.780. At from 180° to 250° C., light or illuminating oils are found, known as photogene or kerosene, the density of which should be 0.800 to 0.820.

The temperature may be gradually raised to 400°, when the heavy oils are obtained, with a density of 0.840, and which are employed as lubricating oils for machinery. Throughout this operation, parafine is distilled and deposited in large subterranean reservoirs, where it is congealed, at a low temperature.

Once it is refined, there remains at the bottom of the retorts a substance called coke, or a residue, which is employed as one of the best classes of fuel known up till the present time.

Five products are obtained by refining petroleum, viz.,

Essence of petroleum,
Kerosene,
Heavy oil,
Parafine, and
Coke.

*Essence of petroleum or naphtha.* This first purified product appears in the form of a transparent colorless liquid, reflecting certain violet hues. It is extremely volatile, and, exposed to the air, evaporates rapidly, without leaving any residue.

By the distillation of essence of petroleum, the following products are obtained:

$1^{st.}$ *Petroleum ether or rigolene.* A substance so volatile that, at a normal temperature, it makes the thermometer fall to 28° 34. It is employed as a dissolvent of resins, and in surgery as a local anæsthetic.

$2^{nd.}$ *Gasoline.* This is employed in the extraction of vegetable oils, the cleaning of wool, and the fabrication of various gases.

$3^{rd.}$ *Benzine.* Benzine is employed to take the grease out of woven materials, to dissolve caoutchouc and gutta-percha, and in the manufacture of certain varnishes,

$4^{th.}$ *Mineral essence.* This substance burns in specially constructed lamps, and is employed in the manufacture of gas, such as air gas and instantaneous gas.

5th. *Artificial turpentine.* A number of colours are obtained from this, and it also replaces essence of turpentine.

*Photogene or kerosene.* This is the second product obtained by the distillation of crude petroleum, and it has to be submitted to a special operation before it can be employed as illuminating oil. It is purified in the following manner: it is first cleansed with plenty of water, and afterwards with dilute sulphuric acid, in order to remove from it all smell, as well as any tar that may have remained behind, which operation is continued until it remains colourless. An alcali is then applied, destined to neutralise the sulphuric acid, when it may safely be delivered to commerce. Nevertheless, some refiners, more scrupulous than the rest, submit it to a further operation, which consists in subjecting it to a high temperature, until the last particle of benzine disappears.

Once the kerosene is refined, which we improperly call petroleum oil, it is absolutely safe, and may be employed with impunity. A lighted match thrown into this oil, so far from igniting the oil, is itself put out. The dangerous accidents attributed to petroleum, should rather be attributed to the mineral essence, or essence of petroleum, which is, as we have shown, the product of the distillation of substances essentially explosive and volatile, as well as to imperfectly refined petroleum, handed over to commerce before separating from it the dangerous products it contains.

It is a very easy mater to ascertain the good quality of the oil. Is should mark on the gauge 0.800, or at least 0.790. If it shows a density inferior to this, its use is dangerous. In the absence of a gauge, a litre of oil may be weighed, and its use will be unattended with danger if it weighs 800 grammes, but not if it weighs less.

The point of ignition of refined petroleum varies between 35° and 45°. Care must be taken not to expose the liquid to a higher temperature because then it recuperates part of its inflammable properties, and may explode on contact with fire.

To prove the safety of refined oil North Americans make the following experiment: they mix a glass of the oil with hot water, at a temperature of 5° C, they then apply a lighted match, and if the petroleum does not ignite on contact with it, the oil is of good quality.

*Heavy oils.* These, which constitute the third product of petroleum during the operation of refining it, are obtained at a temperature of 280° to 400°, and may be classified as follows:

1$^{st.}$ *Solar*, an inferior quality of kerosene, with a density of 0.820. It has a strong smell, a darker color, and gives an opaque light. This oil is obtained by submitting the heavy oil to distillation like crude petroleum. It yields 50°$_{o}$.

2$^{nd.}$ *Parafine* or *Belmontine*. Parafine in a liquid state, is a transparent and colorless substance, and in a solid state, it resembles alabaster in whiteness. It liquifies at 50°, evaporises at 400° undergoing partial decomposition. Soluble in ether, benzine and sulphide of carbon, it mixes perfectly with stearine, sperm, wax and resins. Extracted in the manner we have detailed, and afterwards congealed at a low temperature, parafine is subjected to great pressure. The liquid which it gives off is a highly appreciated lubricant, while parafine remains in the press in the form of a dry white block.

England imports immense quantities from Rangoon in order to extract parafine oil, employing the residue in the manufacture of extremely fine candles.

In Caucasia parafine is employed preferently, it being distilled into large receptacles, and giving 60% of parafine and 8% of oil.

*Parafine*, called crude parafine, reheated with naphtha, and submitted to a pressure of three atmospheres, gives a crystalised product, soluble at 56°, of little value. It is sold in India as chewing gum, on account of the use it is put to by the natives of the country, a single merchant in Bangor distilling 75,000 pounds per annum.

*Parafine* has been given many industrial applications.

Apart from what is extracted from refined petroleum, there is another kind, distilled in the bowels of the earth, and called *ozokesite*, a kind of mineral wax, found in the vicinity of the petroleum fields of Galicia, near Bovislaw, where it is worked on a vast scale, as an article of much importance.

Ozokesite has the yellow color of honey, is transparent, and has the same consistency as honey. There is another kind, of a blackish hue, containing a great deal of petroleum. The formation of this substance is attributed to the oxidation and condensation of hydro-carbonides of petroleum

Melted and separated from the earthy matters it contains, ozokesite is used in the manufacture of artificial wax, employed, above all, in Russia, in the the manufacture of candles

Ozokesite when distilled, gives:

  A. *Benzine*,  2 to 8 %
  B, *Naphtha*,  15 to 20 %
  C, *Heavy oil*,  15 to 20 %
  D. *Coke*,  10 to 20 %
  E. *Parafine*,  36 to 50 %

3rd. *Vaseline.* This new substance was discovered in North America in 1873, and was sold under the name of *Petroleum Jelly*; it was under this name that it figured in the great Exhibition of 1876.

*Vaseline* is a pale yellow transparent substance, of the consistency of gelatine. It has no taste nor color, nor does it volatilise at an ordinary temperature. It melts at 35°C. Its density is 0.840. It is soluble in water, and little soluble in alcohol, although it dissolves in ether immediately. In the liquid state, it mixes with greasy substances, oils and glycerine. It is obtained by evaporating, in the air, the heavy oils which are the result of the refinement of petroleum, and which still contain some parafine.

Some people confound vaseline with *cosmine*, both being of common origin.

4th. *Lubricating oil.* This is a thick and greasy substance, similar to tar. It is generally employed to lubricate machinery, and its use, on account of the good results it has given, has become general.

5th. *Tar.* When the process of distillation of the petroleum is finished, there remains, at the bottom of the retort, a kind of tar, from which another and very fine lubricating oil can be obtained. This tar is given various applications, such as fuel, varnish for walls (preserving them from damp) and also as a powerful disinfectant.

6th. *Coke.* This is the name given to the solid product remaining after the distillation is completed, and is the residue left after the manufacture of petroleum gas. It is the best fuel known.

TABLE SHOWING THE SPECIFIC GRAVITY AND CALORIC POWER OF PETROLEUM IN THE THREE PETROLEUM REGIONS OF THE WORLD.

| Locality of the fields. | Specific Gravity. | Caloric Power. |
|---|---|---|
| Perú (Piura) | 8.480 | 13.672 |

*United States and other Places.*

| | | |
|---|---|---|
| White Oak | 0.873 | 10,180 |
| Burning Springs | 8.412 | 10,213 |
| Oil Creek | 0.816 | 9.963 |
| Ohio | 0.887 | 10.399 |
| Franklin | 0.886 | 10.672 |
| Pennsylvania | 0.820 | 8.771 |
| Heavy Gas Assoiation | 1.044 | 8.916 |
| Parma | 0.786 | 10.121 |
| Java Daudang | 0.923 | 10.831 |
| Jarat Jibodas | 0.823 | 9.593 |
| Java Communo Gogor | 0.972 | 10.183 |
| Bechelbroin | 0.912 | 9.708 |
| " crudo | 0.892 | 10.020 |
| Schual Weiller | 0.861 | 10.458 |
| Galicia | 0.870 | 10.005 |
| From West Galicia | 0.885 | 10.231 |
| Vagnas | 0.911 | 9.046 |
| Autun | 0.870 | 9.950 |
| Mount Marjan | 0.985 | 10.081 |

*Russia.*

| | | |
|---|---|---|
| Bakhu | 0.882 | 11.370 |
| Caucasia | 0,928 | 11.000 |
| Transcaspian Region | 0.987 | 11.060 |
| " | 0.884 | 11.660 |
| " | 0.938 | 11.200 |

The proportion of kerosene found in the petroleum of the three petroleum regions of the world is as follows:

| Peru. | North America. | Russia |
|---|---|---|
| 65 to 70 % | 70 to 75 % | 25 to 30 % |

In medicine and pharmacy, petroleum has many important applications, more especially in homœopathy.

It is employed, with magnificent results, in cutaneous affections. In cases of quinsy, the throat is painted with it. It rapidly heals all wounds, and, taken in doses of one drop, is excellent for asthma and chronic bronchitis. Taken in greater doses it is a violent poison. Its external application for the relief of rheumatic pains is attended with good results, and it is, in general, applicable to all complaints for which phenic acid is receipted.

In pharmacy it is known as cosmeline, vaseline and petroline.

With regard to its applications in other branches of industry, these are very numerous, especially as a caloric force.

A North American engineer has conceived the idea of employing it, with an apparatus of his own invention, as a means of coast defense, setting large quantities of it on fire on the surface of the sea, and through which it would be impossible for the enemy's ships to sail

# PART THE FOURTH.

## CHAPTER I.

*Combustible Oil.*

As we did not dwell extensively enough, in the third chapter of the first part of this book, on this new fuel, which is the most important application petroleum has received, we will enter into a brief study of it here.

The name of *coke* is given to the *solid* matter that remains in the bottom of the retorts, as we have already said in the last chapter, once the distillation of petroleum is completely finished. Coke was considered, for a long time, completely useless, and in the first refineries which were established, it was thrown away, as a product of no value nor importance.

Once tried as fuel, it proved to be the best fuel

known, and the results obtained were so satisfactory, that its use became general, in all places where petroleum was found, as fuel for steam engines.

From this important discovery it was deduced that not only the *solid matter*, but also the oil in a liquid state, might be employed with good results as fuel for all steam engines.

France, England and North America were greatly preoccupied with the subject, in their endeavors to arrive at a satisfactory solution of the problem. Different commissions were entrusted with its study, and, after making many experiments, they decided that, employed in equal quantities, petroleum gave three times as much heat as coal. The French commission, in their experiments, obtained in 17 minutes, with the combustion of 1.92 kilogrammes of petroleum, a steam pressure equal to that produced in 30 minutes by 4.23 kilogrammes of coal. The fire was brought to its greatest intensity in a minute and a half, and was extinguished in the same length of time.

It was then that M. Saint-Claire-Deville, in company with M. Dieudonné, made several experiments with the locomotives of the state railways, and, seconded by M. Depuy of Lome, repeated them on board a man-of-war. The results showed that coal could be replaced with petroleum. This innovation met with a certain amount of resistance on the part of scientific men, who did not look upon any of the experiments seriously.

At that time, the employment of petroleum as fuel looked so impracticable, that one of the men who have written most about American industries, and especially about petroleum, Mr. Simonin, acknowledging the great advantages of the oil, affirmed: "it is only in very special cases that it may be predicted that petroleum will one day advantageous-supplant coal."

At the same time (1869), a German mechanic made some experiments in a refinery at Bakhu. After employing different apparatus invented by him, he eventually adopted a kind of sieve, through which the petroleum passed and was slowly burnt.

In the meantime, the Russians, who wanted to utilise petroleum for their men-of-war and merchant marine, under the influence and direction of the admiralty and a scientific committee, ordered a few government engineers to study the subject seriously, and they shortly afterwards submitted several important projects.

The object sought to be attained was to find an apparatus something like a pulverizer, that would throw the petroleum into the furnaces in the form of rain or spray.

Months after the commission was appointed, an Englishman and a Russian disputed the honor of the discovery, and, although the new apparatus was a step in advance, the problem was not considered solved from a practinal point of view.

Kamenski, a Russian government engineer in Bakhu, secretly obtained the plans of the apparatus that M. Saint-Claire Deville had invented a year before, and successfully applied it to an imperial yacht of 70 horse-power. Kamenski modified the apparatus somewhat, and its invention was attributed to him; but it did not give satisfactory results.

In the face of these fruitless attempts, the Russian steamship companies sent engineers to the principal European centres, where it was eagerly endeavoured to solve the petroleum problem. Lenz went to London, where he put himself in contact with Aydon, the inventor of an apparatus, and afterwards, in Paris, with Mr. Deville. The French apparatus was adopted by Lenz, who believed to be the more perfect of the two, but, when it was tried by himself on board the *Derjavin*, it did not work at all. He then modified it on Aydon's plan, and obtained the best results.

It was in 1872 that Lenz, after making many modifications in the apparatus, too long to enumerate, succeeded in inventing one that satisfactorily solved the problem. and this is the apparatus used since that time, and employed to-day, in all the steamers that sail in Russian seas and rivers.

Having achieved this great success, Lenz endeavoured to apply the apparatus to locomotives on railways; but the result was not complete, for it was claimed that his system deteriorated the tubes and did not distribute the heat equally. Engineer Brand then invented a new apparatus without this grave inconvenience, which is the same as is used in all Russian locomotives, as well as on locomotives of the Arequipa, Puno and Cuzco, and Central Railways in Peru.

Theoretically, a furnace consuming petroleum should consume a ton of liquid to two of coal; but, in practice, in well constructed furnaces, the economy is much greater, and may be calculated as one to three. The result therefore cannot be more satisfactory, looking, in the first place, at the reduced bulk, and, in the second place, at the economy in the cost. In France, for example, according to M. Hué, a ton of residue is worth ten francs, while a ton of coal costs thirty.

## CHAPTER II.

*Conference in London, February 13th, 1891*

For the better information of readers on this important subject, we give, entire, the speeches made at the conference in London, in 1891, by Captain Coxmichael and the Chairman of the conference, both of which we take from the London *Shipping Gazette and Lloyd's List.*

Captain Carmichael said:

The subject of liquid fuel in steamers is not a new one; boats have been run with it on the Caspian for several years. In 1876 a boat called Supe was run with it on the coast of Peru, under the superintendence of Mr. Bryce Douglas, of the N. A, and C. Co. Boats have run with it in England, on the Thames the Wear, and other rivers; but, partly owing to the timidity of owners to adopt it, the use of oil fuel has not been much taken up by steamship owners. Now, however, the aspect of affairs is changing. Steamer owners that formerly were either owners of sailing ships, retired captains, or businees men pure and simple, are nowadays men who acquire a technical knowledge of the working of their steamers in all departments, and it is generally a bad speculation for either a captain or engineer to try and persuade them they don't know all about it. I don't intend to waste your time by a long introductory address on general subjects, as I do not wish you to go to sleep. I mean however, to try and explain to you, as shortly as possible, how petroleum is used for fuel; how I have seen it fired in an ocean-going steamer for many months, and why I advocate it so atrongly; also my own personal experience and observation, while in command of a steamer burning it: after which I propose to ilustrate its usefulness from a commercial point of view, together with its advantages over coal for torpedo-boats, and men-of-war generally; and where obtainable, its advantages for merchant steamers, especially for passenger boats. First of all, therefore, I shall describe the fuel used so far with perfect success, after which I shall describe the method of using it. In Russia it is called "Astacki"; in England and America it is generally known as "Residuum". It is the residue of crude petroleum when all the volatile or lighter oils are distilled from it: it is perfectly non-inflammable until heated to 350 deg. and consequently quite safe to use and carry in large quantities; it has no smell, in consequence of not giving off gas, until obtaining the required heat; it does not deteriorate by being stored in tanks, exposed to the air, nor does it evaporate perceptibly; it is not detrimental to metal tanks, the inside shin of the vessel nor any of the ordinary receptacles.

### *Method of using it.*

Its advantages from an engineering point. The fires are completely under control of the engineer on watch; he can regulate them so as to have any pressure of steam he wants without being dependent on his firemen; in heavy weather his steam is as much under control as in fine weather; it does not need any extraordinary attention, as when once the burners are set away properly, they will run without any trouble for days, and maintain steam almost without variation, providing the engines are kept in the same gear, and feed water is used in the regular way. This I have proved on several voyages during the last eight months, notably on my first voyage from Talara Bay, Peru, to Valparaiso, Chili, in the steamship Ewo; the pressure on the boilers was maintained steadily at 158 lb. for seven days, and under most unfavourable circumstances, as in the first place it was our first ocean trip with oil fuel, and the ship was being driven full speed into a head sea all the time. For the information of engineers, of whom I see a rather formidable array, I must here state that during that time, in order to compensate for the loss of fresh water occasioned by the use of the steam jet, together with the ordinary leakage from glands, &c., the boilers used ten tons of fresh water; the boilers being run up in Talara Bay, showed on being tested in Valparaiso, after the seven days' run, a density of 7 cz. of salt; the engines developed 450 i.h.p. The contraction and expansion of the boilers by using oil fuel is reduced to a minimun as in getting up steam you can do so as gradually as you wish. There are no firedoors to open every time you coal your furnaces; your fires never need cleaning: consequently, that fruitful cause of loss of steam, and contraction of the boiler is done away with, and not having to force your fires to regain what was lost in that operation you avoid the consequent expansion of the boiler, thus the life of a boiler fired with liquid fuel must from these circumstances alone be longer than a boiler fired with wood or coal. Liquid fuel, if applied with steam, does not injury to the crown sheets, back conections, or combustion chamber of a boiler; tubes do not need sweeping often—as with coal, for instance; the tubes in the steamer I mentioned were only swept once in six months, and even then were not dirty; there is no dust or grit, as there are no ashes; therefore, the saving of labour and paint are considerable items in the engineers' department.

*From the shipowner's point of view.*

Machinery that is kept clean will run longer than machinery that is grinding grit. Boilers that are always at the same temperature will last longer than boilers that are subject to frequent variations of temperature; therefore both machinery and boilers will last longer using oil fuel than using coal. Firemen's wages, in these days of Trades'Unions, are somewhat of a grief to shipowners, and we hear of many steamers being detained very unnecessarily owing to the peculiar laws at present in force regarding both seamen and firemen. If you can work a steamer of 3,000 tons with only three firemen, it should make the shipowner happy, but the great advantage gained by the shipowner is that he can carry his fuel in a space that is now wasted, namely, the cellular bottom of the ship, or in ballast tanks, the consequent carrying capacity of the ship being increased by the space which was formerly taken up by coal. As the consumption of oil, as compared with coal, is (weight for weight) one half, or in other words ten tons of coal are equal to five of oil (these facts are taken from actual experience), it is obvious that the storage of oil fuel can be made much more compact and easy of access; the ship is more speedily fuelled and you avoid port charges, by being enabled to carry more fuel in proportion than formerly. As an instance, we will go into figures; say, a steamer of 2,200 tons gross register, with 750 i.h.p. uses 16 tons of coal per 24 hours on a voyage of 105 days actual steaming, would consume 1,680 tons of coal, against 800 tons of oil. The cost of the coal at, say, 20s. a ton, for both outward and homeward passage, which is rather a low average, would be 1680*l*., while 800 tons of oil at 40s. a ton, which is rather a high average, would be 1,600*l*. The wages of six stokers for coal would be 112*l*., against the wages of three stokers for oil, 56*l*.; cargo space gained 1.050 tons, which at 40s. a ton would be a nett gain of 2,100, or a nett saving of 2,236*l*. per 105 days, or, roughly speaking 7,000*l*. per annum Owners are, I know, sceptical as to figures and theory generally, and will very naturally ask—Where are we to get this fuel from? And if we make the demand for it, will not the producers put the price up? To these questions I would answer, that as petroleum is known to exist in all parts of the habitable globe, and as the laws of demand and supply remain unaltered, it is hardly probable that competition would not bring this fuel within easy reach of the commercial world. Experience has taught us that both enterprise and capital were never wanting where the capitalist can see a sound investment. Of course there are certain expenses entailed by the use of oil fuel; for instance, the furnaces would have to be fittted for burning oil, the average expense being about 30*l*. per furnace. You would have to make your side bunkers oil tight, and fit them with the necessary connections, also some alterations might have to be made about the ballast tank, a small steam pump would have to be fitted, together with a supply tank, and the necessary connections for steam and oil. These expenses would be done away with entirely, if the boat were originally built for oil fuel, as in the boat I have just commanded.

*Its use for naval purposes.*

For many reasons oil fuel is of more value to

torpedo boats and ships of war than it even is to merchant steamers, being capable of such compact storage; for you are able to utilise parts of the ship that are now useless, with the additional advantage of keping your fuel completely out of gun fire. Again, a most important question arises, and one which was asked in Parliament, but if I remember rightly, very ably shuffled out of. "How are we to fuel our-men-of-war at sea during a blockade?" To either a marine engineer or a seaman the importance of this question will present itself. Last summer's naval manœuvres went to prove the difficulty, for although, under exceptionally favourable circumstances, men-of-war were coaled at sea, the Admiralty and all the naval officers who witnessed the operaration know that this means of obtaining fuel could not be depended on, and, besides, the modern man-of-war is not constructed for rapid coaling. One of two things would have to be done in the event of a blockade; either the ships would have to be relieved, or else they would have to be fuelled at sea. If the former were adopted it would mean more ships and more men; and if the latter it would mean more coals, more men, more boats, and more fine weather than we generally get. The reason I say more coals is because no mention has been made of the coals which went overboard when coaling at sea was tried. The reason I mention more boats and men is that I have always noticed that when such operations are carried on in a seaway boats get unaccountably damaged and have to lay up for repairs, and it has generally been my experience also that a certain number of men also needed repairs about that time. Now with oil fuel the case is very different; the large tank steamers of the present day could always fuel the fleet, even in heavy weather.

For instance, if I were going to fuel a man-of-war in bad weather, I should signal her to throw a rocket line aboard me, and haul a hawser aboard from her stern to my bows, in the usual manner described by the Board of Trade rocket apparatus. Having the hawser aboard and made fast, I should signal the war-ship, "steam ahead slow," and then let her send aboard the tank on the hawser a line, to that I secure my hose, which the man-of-war can heave in, and when it is coupled up, I pump in the fuel at any rate I wish, according to size of hose and power of pump.

The hose, of course, should be secured to the hawser by hanks, which would enable it to be kept straight, besides being a great help in getting it aboard again. Common canvas hose will do, as residuum will not penetrate it, especially the seamless, double-ply woven hose. Of course, in fine weather the operation could be made much more simple and rapid, as you could have two or more hose employed. The question was asked me by the captain of the Chilian iron-clad Blanco Encalada, "What would be the result if a live shell were to penetrate a fuel tank, and explode?" I told him I beleived it would be a bad job for the shell, as it would put in out; but granted that it did explode. it would not do as much harm as if it had exploded in a coal bunker, because, in the event of a shell exploding in the said coal bunker, each piece of coal becomes a projectile; and if any men were in the coal bunker at the time, it would make it lively for the men. With oil, on the contrary, it would produce the same effect as if it exploded in a tank of water, as the volume could not be brought up to the required temperature for explosion; besides, a man-of-war or torpedo boat constructed to use oil fuel would have her tanks completely out of gun-fire, with the advantage of being able to utilise the space now occupied by coal bunkers for other purposes. Again, the heat in the stoke-hole of an oil steamer is reduced to a minimum, and closed stoke-holes and fixed draught need not be mentioned where oil is talked of as you have all the forced draught you need with the steam jet and oil fuel. Again, men-of-war of the present are such masses of machinery that there is very little room left for coal or crew space. As the consumption of oil is, weight for weight, one-half that of coal, men-of-war increase their efficiency as much again by the use of oil fuel, while to the crew it means more space and less dirt. The waste with oil fuel is not so great as with coal. Engineers know full well the waste there is with coal, in having to drive ahead full speed at short notice, with green fires, and then having to suddenly "stop", as in an action or during manœuvres; whereas with oil the fires are as completely under control as if you had a gas jet to handle. I believe the Admiralty are thinking of burning oil fuel, but have not as yed decided what kind to use. I should certainly say not crude oil, but residuun, for a hundred good reasons, of which to-night I shall only state a few. 1. Its being so perfectly safe while on board ship. 2. It can be stored on shore so easily, and, in the event of a fuelling port being bombarded, shells could not set fire to the tanks. 3. It can be handled without risk of fire or explosion from the shore to the ships, or *vice-versa*, and it does not deteriorate by exposure to the weather. In conclusion, I must beg your indulgence for the crudeness of this paper. It would not be possible to go into the subject otherwise than in a superficial way in a single night, but my object is to allow the mercantile world in general, and the members of this society particularly, to know the progress oil fuel is making, the comfort it is to engineers and all concerned, its pecuniary advantages to the merchant who can avail himself of it, its peculiar adaptability to ships of war, and its special value to large first-class passengers steamers where cleanliness and the comfort of the passenger, is an object. If I have been able to arouse your interest in this subject, and have made myself clear, I feel amply repaid, and thank you very sincerely for your patient hearing.

A long discussion followed.

The CHAIRMAN said the subject of the paper was one of the greatest interest to their profession, and to commerce in general. When they saw how the supply of coal was being exhausted, they began to recognise the necessity of getting some other fuel. Captain Carmichael's paper had been written to open discussion on the subject, and he invited any gentleman who felt himself competent either to adress the meeting or to ask the lecturer questions. He addressed this invitation to visitors as well as to members of the society.

Colonel HARRIS, of the Petroleum Company, said that in 1882 the price of coal on the west coast of South America was 3$l$. per ton, and during a small trial there was a saving of two-thirds of the combustible matter by means of petroleum. The Steam Navigation Company would at that time have altered all their furnaces and used oil could they have depended upon a regular supply of oil; but the experiment, which was extremely valuable in showing the saving effected by the use of oil as compared with coal, dropped. Within the last three or four years new petroleum wells had been opened, and as Captain Carmichael would tell them, there was any quantity of oil they chose to take away. Six weeks ago he had a letter, stating that one well 18 inches in diameter was throwing a stream of oil 90 feet high, and experts said that the quantity of oil on the south-west coast of Peru was equal to any quantity which had been obtained from Valparaiso. It was very important that some substitute should be found for coal, and the paper read by Captain Carmichael would show that oil had only to be tried to be approved. The British Government were, however, very slow in their movements in all these things, and he thought the mercantile marine would be the first to take oil up as a fuel. If they could get oil at a reasonable price, they might save an immense sum, not only in regard to coal, but in other matters—in stokage, in the wear and tear of journals, and so forth. In fact the saving was so large that it would take a long time to explain it. He had looked at the burning apparatus shown by Captain Carmichael, and, in his opinion, it was the most perfect system for burning oil of which he knew at the present moment. Captain Carmichael's paper was valuable not only to the merchant service but to the Admiralty, who wanted rousing up. He thought they were burning oil on the British men-of-war in the Black Sea at the present time, and he believed it answered the purpose very well. There were many oil centres now being opened. One for instance, existed on the West Coast of South America, where coal was 3*l.* per ton at contract price. There oil could be put into the ships alongside the wells at 1*l.* per ton, and has one ton of oil would do the work of 2 tons to 2½ tons of coal, he thought oil would gradually force its way into use. As to the supply, he would guarantee to put as much into their steamers as was wanted. He would supply 100,000 tons per year at 1*l.* per ton. Oil might also be used in the manufacture of gas, and where it could be obtained at a low price, it was very much superior to coal for this purpose because it made an illuminant very much superior to coal gas. A ton of kerosene would contain more than three times the cubic feet of gas the coal would contain, and he believed that gas works all along the coast would consume oil. On the west coast of South America the steamships of the Chilian Company and the South Pacific Company consumed upwards of 100,000 tons of oil a year, and they could get as much power out of 100,000 tons of oil as out of 200,000 tons of coal while oil was only one-third the price of coal. In 1887 no less than 460,000 tons of kerosene was exported from New York to Oriental countries. He thought the meeting was greatly indebted to Captain Carmichael for the valuable information he had given on the subject —(hear, hear).

Mr. E. HENWOOD informed the meeting, that, as an engineer, he had had a good many years, experience in the use of liquid fuel. He began in 1883 to experiment in his own yachts. He had to be content with green oils, with creosote which was rather heavier in specific gravity than the oil they got from Rusia, but he would prefer Astacki, which was to be had in greater quantities. Subsequently he tried it in a larger vessel, and in 1886 explained his system to a great number of superintending engineers and others. The Admiralty also sent their engineer to see it, and they saw the evaporation he was able to effect by employing super-heated steam. By this system the consumption of coal per engine horse-power was reduced from 2 lbs. to $\frac{3}{4}$ lb., and in triple engines to something less. He afterwards fitted the system to his own launch, which had a return tubular boiler.

The vessel was seen by Lloyd's surveyor, and a very favourable letter was written regarding it by their chief surveyor. The Admiralty surveyor was on board for four hours, and was perfectly satisfied. There was not one trace of smoke the whole distance; they had no difficulty whatever in keeping steam; and he could dispense entirely with the services of the stoker. In fact he sometimes had dispensed with the stoker, having only a waterman on board to steer. The paper was excellent throughout, and a more interesting contribution to the records of the society could not be made. He was in perfect accord with nearly all Captain Carmichael had said, but he preferred his own injector to that shown by Captain Carmichael, because it atomised the fuel in a more effective way, was far lighter, and could be more readily disconnected. In ships using oil every stoker might be dispensed with, and in their place a third-class engineer may be emgaged. These men were of rather better education than the stokers, and all they would have to do was to keep the oil tanks charged at the proper time. He did not remember whether Captain Carmichael mentioned the absence of smoke or smell in the oil ships. One of the directors of the Union Company once went on board his little vessel "Ruby," and was quite astonished when he found neither smell nor smoke.

Mr. JAMES HOLDEN, of the Great Eastern Railway Company, thoroughly endorsed all that Captain Carmichael had said as to the value of liquid fuel. He had been very interested in the paper, and he was among those who for years had been surprised that the mercantile marine and the navy had not taken up the subject of liquid fuel very much more seriously than they had done. On the Great Eastern

Railway, in whose service he was employed, they had had at work for several years past a number of locomotive engines which used liquid fuel, and last year one of those engines travelled 47,000 miles without a single failure or a single difficulty. It was an express engine running between London and Norwich, and the apparatus was so easily controllable, so perfectly safe, and there was such an entire absence of noise or smell that 99 passengers out of every 100 knew nothing about the engine being propelled by liquid fuel instead of coal. One of the advantages which he claimed for the system he used, was that he had not to alter the furnaces in any way to use liquid fuel. On one journey he could use liquid fuel and on another solid. He could even use liquid fuel one minute and the next solid, without any alteration of the furnaces, and in the present tentative state of the subject, that appeared to him to be a considerable advantage. The difficulty of getting an assured supply of liquid fuel would make a shipowner reluctant to alter his furnaces so that they could use liquid fuel alone. His own difficulty in extending the use of liquid fuel had been this difficulty of getting the fuel. The Great Eastern Railway Company had at present six engines using oil. He had made extensive inquiries with the view of getting a regular supply of liquid fuel at a reasonable rate. He had inquired in England, in Russia, in the United States, and in Canada, but he could not get an assured supply over a term at anything like reasonable rates. Therefore, if that discussion led to some action on the part of gentlemen who were interested in the supply of liquid fuel, a very great step would be gained. He had some 200 liquid fuel injectors at work in all parts of the world—in India, and the West Indies, South America, North

America, Scotland, various parts of England, but the difficulty in every place had been that of getting an assured supply of fuel. With regard to the use of Astacki, he had been particularly using the tar from the manufacture of heavy gas for the lighting of railway carriages; and on the Great Eastern Railway they had some 1,000 carriages lighted with gas made from shale oil. They got the tar from that because it was not readily saleable, and they used it to propel their locomotives. They had also used ordinary tar obtained from the gas works, as well as green oil—in fact, anything they could get at a reasonable price. He had used Paramatta oil and Trinidad oil, and he found that Paramatta, or Trinidad, or Astacki oil had very much more calorific value than the materials which they had been compelled to use, owing to the difficulty of "residuum" at a reasonable price. The Great Eastern Railway Company had a number of stationary boilers in use at their works, fed by liquid fuel; and a ferry-boat, running on the Thames between North and South Woolwich, was fired by liquid fuel. In none of these cases had the slightest difficulty ever been experienced, with the exception of getting a supply of the fuel, and that was the principal difficulty likely to be experienced. He apprehended that the officers of the Board of Trade would probably be willing to allow a very much more extended use in regard to Astacki; but he did not wish at all to criticise their very careful way of going into a question of this kind. It was right they should be very careful; but with the use of ordinary "residuum" there was an entire absence of danger, because, as Captain Carmichael had pointed out, it was non-explosive; and with ordinary care there was no more danger than with coal. On board ships it might be used either with force or natural draught. With regard to the loss of fresh water, he should have thought there would be no difficulty in using a small donkey boiler for getting a steam supply.

Mr. HENWOOD wished to add that he found no injury resulting to the boilers from the use of oil, and it was impossible to set fire to the oil even with a red hot iron. So far as the supply of this oil was concerned, he was now negotiating for establishing depots in different parts of the world, and he believed there would be an immense supply before long. There were wells in Burmah, in addition to the other places mentioned, and a Dutch firm with a capital of £100,000 had been formed for the purpose of working oil in Java.

CAPTAIN WILSON BARKER concurred as to the great difficulty of getting liquid fuel. He was very much surprised to hear that it had been used in England, but in Russia it had been used with very great success for many years. They found that the air blast did not work at all well; the heating of the under surface of the boiler was very intense, and in many cases the rivets were actually melted off the boiler; but when the steam blast was used everything went on satisfactory. He was surprised to hear Captain Carmichael say air did not affect the oil. He should have thought it would deteriorate when brought in contact with air.

CAPTAIN CARMICHAEL, interposing, said that in Peru the oil tanks covered a tremendous area. One tank alone held 37,000 barrels of oil; and the whole of that oil was left exposed to all the winds that blew. It had been so exposed for over a year, and, so far as could be seen, there had been no evaporation and no deterioration, the oil being in exactly the same condition as when it was put into the tanks.

The coast of Peru seemed to have a temperature specially adapted for this oil, and though the fact was not mentioned in any of the sailing directories, he had noticed the Artic current, which seem to account for the singularly even temperature—a temperature which at night went down to 68 degrees, and in the daytime never rose beyond 78 degrees.

Captain Barker said the current to which Captain Carmichael referred was well known, and there was a similar current on the West Coast of Africa.

Captain Harris also remarked that he had many times traced the Artic currents off the coast of Peru, which was one of the most delightful countries to which an old man could go.

Lieut. Pybus, R.N.R. The subject introduced in Captain Carmichael's admirable paper points to changes of vast interest to the shipowner, shipmaster, and ship's officers, and I have taken the trouble to verify some of his statements. It is necessary to consider the various points in connection with the use of oil to enable us to grasp all the details in connection with it and to ascertain if it is used in the most practical, and, secondly. in the most economical manner. The fuel advocated has another merit; that is, it is the residuum after all the most valuable constituents have been extracted. The crucial point, and the one which I think will meet with most opposition from an engineer's point of view, is the question of using the steam jet instead of air. It certainly has the merit of having stood the test of practical experience, as it is largely used in America. To consider it properly, we must go into the question of combustion. What is combustion? It is simply the chemical combination of two gases to form a third, by which great heat is evolved. In an ordinary coal fire the oxygen of the air enters into combination with the carbon of the coal to form carbonic acid gas. In the present instance we must consider our elements of combustion. In coal we have a varying proportion of carbon (depending on the quality of coal combined with gas) composed of hydrogen and carbon. In petroleum we have 85 per cent of carbon with 15 per cent of hydrogen. In steam we have two parts hydrogen with one of oxygen, or 89 parts oxygen with 11 parts hydrogen to form 100 parts water. In air by volume we have 21 parts oxygen, 79 parts nitrogen to form 100 parts air. For the complete combustion of marsh gas (one of the hydro carbons and similar in its constitutents to petroleum), and one of the chief ingredients of coal, you require a double quantity of oxygen or 10 volumes of air as one half the volume of oxygen, combines with the carbon to form carbonic acid gas $C_2$ and the other half combines with the hydrogen and forms water. For the complete combustion of a pound of carbon it requires 12 cubic feet of oxygen. In the air this gas is diluted with four times its bulk of nitrogen. In theory 62 cubic feet of air is therefore required to supply the necessary quantity of oxygen; but all the oxygen is not extracted from the air as it passes through the furnace, so that in practice 150 feet of air is required for each pound of carbon consumed, the exact quantity varying with the shape and construction of furnace. If too little is admitted the combustion will be imperfect, if too much the temperature will be reduced, causing a large waste of energy. In using petroleum we have a substance containing a large proportion of hydrogen. Hydrogen has a greater affinity for oxygen than for carbon, so that its first action would be to seize 89 parts of the oxygen and form water, leaving very little for the combustion of the carbon. If we use steam we at once supply a large amount of oxygen, combined with hydrogen, which is in itself an element of combustion. I have not a sufficient knowledge of chemistry to be able to indicate what the new combustion will be, but one thing is certain—that in combining they will evolve heat. I only wish to point out that in every volume of water we supply in the form of steam e wsupply 89 volumes of oxygen necessary for the combustion of the carbon, and in every volume of air we only get 21 parts oxygen. Oxygen constitutes 1-5th of the volume of the atmosphere, and 8-9ths of weight of water. I also wish to point out that in using air the immense volume you have to use to get the necessary oxygen causes loss in other ways by reducing the temperature. I was much struck by a remark of Captain Carmichael's. When forced draught was suggested he said the intensity of the light from the furnaces blinded the firemen. That indicated imperfect combustion, as the luminosity of flame is caused by the incandescent and consequently unconsumed particles of carbon, which, where the temperature is reduced, becomes visible as smoke. That heat of a flame does not depend on its luminosity is shown in the oxy-hydrogen blow pipe and an ordinary coke fire. I think I have said sufficient to prove that the use of steam in the furnace will more than compensate for the loss of steam in the boiler. I would also remind you that the large amount of latent heat in the steam is utilised and used over again. To meet practical difficulties, I would suggest a small auxiliary boiler to drive the furnace, or the donkey boiler might be utilised when not required for condensing purposes, &c. We now come to the economical aspects of the case. We will first take its evaporating qualities. 1 lb. of Astacki is equal to 1·37 lb. of bituminous coal, and 1·60 lb. of anthracite. In actual practice 1 lb. of Astacki is equal to 1 8-10 lb. coal; space—1 ton of coal is equal to 40 cubic feet; 1 ton of Astacki is equal to 39 cubic feet; weight, one-half. On the whole question the balance is largely in favour of liquid fuel. Captain Carmichael has also given us the cost, which is on the right side. One of the greatest points in its favour is that it can be adapted in any boiler without a large outlay, and the boiler can also be used for burning coal if required, although I think that when oil comes into general use and supplies are available at every port it will cause many modifications in the construction of both boilers and ships. In boilers, by the adoption of the regenerative system similar to that employed in the water gas system for the reduction of metals, by which a saving of from 20 to 30 per cent of fuel effected. In France, at one of the ironworks, a contrivance was introduced for combining hot air and superheated steam to puddling furnaces. The grate, fire-boxes, &c., were connected with air chambers which supplied vapour heated to a temperature of 450 deg. to 500 deg., and superheated steam was applied under the grate, by which a most important saving was effected—in ships by the abolition of large funnels, and open stoke holds, with the attendant danger, dirt, large engine-room staff, &c., which will give greater comfort, more space, and better discipline. The deleterious effects of sulphur from the coal will be avoided and add considerably to the life of the boiler, and the interior structure of the ship.

CAPTAIN CARMICHAEL, in answer to a question in reference to oil wells in the sea, said there was a group of islands on the road to Callao, and 30 miles

S.S.W., where ships always run through a regular drift of petroleum, extending for two or three miles, Sailors could smell it as soon as they got up to it, and the sea at the spot was always as smooth as glass. The oil came from an everlasting well in the sea, which was mentioned by Pizarro, the conqueror of Peru.

CAPTAIN WILSON said there were plenty of oil wells to be found in the sea everywhere. In cable work he had come across a good many of them.

COL. HARRIS added that in some places the oil also oozed from the cliffs.

MR. MATTHEWS, as an engineer, thanked the society for having given him the opportunity of being present that evenig to listen to Captain Carmichael's paper.

CAPTAIN FROUD the secretary of the society, said that air could be controlled just as readily as steam. He could not imagine that we had reached finality in the shape of boilers, nor could he imagine that air would not be used to assist in the combustion of petroleum. He illustrated this process by means of a Wanzer lamp using forced draught, showing the greater heat developed by the assistance of the blow-pipe, the experiment being made in connection with the melting of metal. He urged that more attention should be paid to the use of air in burning mineral oils, and hoped to see some means introduced to enable the oil to be used in perfect safety when brought straight from the tanks.

MR. HOLDEN said that he always made use of a variable and adjustable air supply in using oil.

CAPTAIN CARMICHAEL considered that the use of oil as fuel was an advantage both to passenger steamers and to vessels of war, especially in enabling the latter to cover their retreat.—Votes of thanks were passed to the author of the paper and the chairman.

# REPORTS.

# REPORT TO THE BOARD OF DIRECTORS OF THE "PERUVIAN PETROLEUM COMPANY."

*New York, February 1886.*

Gentlemen :

The magnitude of your Oil Territory in South America, the increasing consumption of refined Petroleum in Peru, Chili and Ecuador, and the proximity to large foreign markets, have induced your Superintendent in his first report to describe more minutely than would have been otherwise necessary, the geographical situation, geological character and natural advantages and disadvantages to be met with in developing the territory and conducting this grand enterprise to success.

These Oil Lands form the foot hills of the Cordillera from Cape Blanco to the low range that constitutes the southern boundary of the Tumbez Valley. They begin at the "quebrada" de Charan, about ten miles south of the Tumbez River, and extend along the coast to a "quebrada" a little south of Cape Blanco—a distance of about forty leagues; averaging in width about twenty leagues; thus giving an area of 800 square leagues equal to 7,200 square miles, or 4,408,000 acres.

The northern portion of this vast tract, where the Company have commenced operations, is easily accessible to Tumbez, which is connected with Paita to the south and Guayaquil and Panama to the north by steamers running three times a month, and by mails taking sixteen days from New York.

*Climate.*

The Company's works at Zorritos (about 20 miles to the south of the Tumbez River) are near the southern margin of the Equatorial Rain Belt.

This great Belt shifts regularly north and south governed by the S. E. trade winds and the variation of the seasons. Its southern margin oscillates between 3° 30' N. and 3° S. causing alternately a rainy and dry season.

At Zorritos it genenrally rains a few nights during the two winter months of January and February. Although at times an interval of two or three years passes with scarcely a shower, on the other hand, the margin of the Rain Belt occasionally extends further to the south enveloping Zorritos for three months in tropical rains.

During the months of December, January, February and March of last year, there were occasional showers at night but the regular operations of the Company were not interrupted by bad weather even for a day.

The temperature is lower than at Guayaquil or even at Tumbez, the thermometer rarely reaching 85° Farht., and usually reading between 59° to 63° through the night, and from 67° to 78° during the day.

Although so near the Equator, the S. E. Trades blowing fresh from the ocean for ten months, and the north-westerly winds for the rest of the year agreeably moderate the temperature and render the climate decidedly healthy.

The Coast Line is tortuous but has a general direction of S. by W. beginning at Punta de Malpelo which forms the southern boundary of Tumbez Bay. From Malpelo Point to Charan is the low, flat coast of Tumbez Valley. At La Cruz, near Charan, there is a point projecting into the sea, another at Mal Paso Grande which forms with Mal Paso Chico an indifferent anchorage, for although the water is deep enough, yet the small bay is somewhat exposed to south-westers, and more so to the north winds.

The next point forms the northern, as Punta de Zorritos does the southern, boundary of the Bay of Zorritos; a good harbour where a ship can lie within a thousand feet of shore, in five fathoms, at low water. Further south the nearest point of importance is Punto de los Picos, at the south of Boca de Pan, forming another good harbour, much frequented by vessels loading fire-wood for the Lima market.

Immediately north of Cape Blanco is the next bay and harbour of importance, that of Máncora, which is well protected and adjacent to Caña Dulce, a locality of much interest to the Company on account of its "Brea" mines and heavy Petroleum.

When your Superintendent first landed at Máncora to examine these mines, he found a ship of a thousand tons in the harbour loading with fire-wood.

This whole region is rough, undulating and hilly. The hills crowd down almost to high water mark, and increase in elevation interspersed with valleys and "quebradas" from the shore to the Cordillera.

Following up the Tuciyal Valley the grade was roughly estimated at one hundred feet to the mile;

this would doubtless increase rapidly after passing the foot hills.

The surface is covered with variegated clays, green sands and fossiliferous conglomerates, consisting of gravel, pebbles and boulders cemented together with lime or ferruginous clay or sand.

From Mal Paso Grande, for more than a league to the south, the sea has encroached upon the hills, unermining them, and forming in many places cliffs of from 60 to 120 feet in height. The strata exposed by these cliffs, as well as by the "quebradas" where the torrent of the rainy season has cut away the hill-sides sometimes to the depth of from 60 to 80 feet, are for the most part variegated clays, red, white, yellow, purple, blue green and brown; blue, red and brown shales; dull coloured friable sand stones, that readily disintegrate; plates and layers of gypsum interspersed between the layers of brown shale; beautiful specimens of selenite and clay iron ore, concretions of argillaceous and ferruginous sand stone; globular and ovoidal, of from one to five feet in diameter (found in the valley of Tuciyal); fossiliferous lime rock, sometimes crystallised, brown coal or lignite, as at Mal Paso Grande, where it crops out in a seam 18 inches thick, between the layers of brown shale bituminous shale exhibiting specks of alum as a white powder, and sometimes layers of what appears to have been beach sand. Mr. A. E. Prentice, Civil Engineer of the Peruvian Government, says in his report upon this section :—

"With the exception of the veins of calcareous "sand stone found near the surface, and an occasion-"al fragment of a large marine bivalve in the grey "sand, very little lime is found until reaching a "depth of from 70 to 80 feet, where it appears to "change into or alternate with strata of very com-

"pact, dry and indurated marl, of the same bluish
"grey colour with specks of carbonate of lime. The
"boring was stopped in this marly shale at 79 ft. 6
"in. below the surface. It was subsequently ascer-
"tained that at Casitas about ten leagues to the
"south (S. E,) in a district presenting the same ex-
"ternal features, a well had been sunk to upwards of
"250 feet in indurated marl of the same description
"which at a depth of 300 feet changed to a very
"compact close grained dark grey (almost black)
"clay slate. The only other circumstances worth men-
"tioning are the briny and brackish water springs,
"found every two or three leagues along the coast,
"A considerable deposit of rock salt which is found
"close to the surface at "Boca de Pan" two leagues
"to the S.W. A few leagues to the northeast of
"Zorritos the hills assume a yellowish white appear-
"ance, and on the tops of some of these I have
"found deposits of conchoidal nodules of horn stone
"or rather opaque, gilicious stone, resembling flint,
"yellowish white on the outside, and dark brown in-
"side, imbedded in a light coloured pulverulent
"marl.

The general appearance of the coast from the sea, as you sail along the shore, is exceedingly barren and forbidding, but this aspect is somewhat relieved on a nearer view by the beautifully variegated colours which the hills present from their argillaceous deposits sometimes yellow then red and green, blue and violet or brown as one or another clay predominates. During the winter season, for several months the valleys are covered with grasses, wild flowers and a rich verdure which creeps far up the hill sides presenting a refreshing and delightful aspect.

No fresh water is found near the shore, from the

Tumbez River to the Chira at Colan, a little to the north of Paita. More to the interior the country is scantily watered, but is susceptible of cultivation in the valleys while the hills and mountains furnish rich pasturage.

Nearly all the valleys and many of the hills in the interior are wooded, indeed, for many years, the "Hacienda de Máncora" has furnished a considerable portion of the fire-wood consumed in Lima.

The most frequent and important tree is a sort of bastard vanilla, the Algarroba, which is in appearance like the honey locust tree of Egypt and the East and also like it produces a very sweet bean, most nutrious and fattening to cattle, horses and other live stock.

The wood is heavier than that of the locust, having almost the gravity of water, it is very hard and of a dark colour.

It is a most excellent fuel, far better than even dry hickory, giving out more heat and lasting longer and is well adapted for steaming purposes.

Delivered alongside the portable engine, it costs six dollars per cord, and there is an abundance of it to develop the entire estate.

Your Superintendent having no facilities for properly examining and classifying the fossils discovered in sinking well No. 3, gives the following geological sketches with some hesitation, and wishes to remark, that he may find it necessary at a future day to correct some of the conclusions expressed therein. It is doubtful whether any epoch of rhe carboniferous period is represented at all, and if not, the tertiary rests upon the upper Devonian.

## Geological Character.

The western chain of the Cordilleras taking the "Vulcan de Sangai," as a point of departure, bends rapidly to the westward, till a little to the S. W. district of Cuenca, it approaches to the Pacific Coast, here less than a hundred miles distant. This chain then bending to the south, follows the coast-line throughout almost the entire extent of Peru and Chili, leaving a narrow strip of lower lands at its base. The geological formation of the Cordilleras is mostly primary granitic, and porphyritic gold is found near Cuenca on both slopes, in the sands of the streams and in the quartz veins. On the surface of the upheaved and almost perpendicular cliffs at Mal Paso Grande, are found beds varying from two to four feet in thickness formed of oyster shells belonging to existing species, agglutinated into a coarse rock. Thin layers of lignites, green sandy marls, and other remains of the eocene epoch of the tertiary period are found at intervals for 60 miles along the coast to the south, and extending inland, as far as has been observed, from two to seven leagues.

Immediately subjacent to these relics of the tertiary period, are found what appear to be shales, sand stones and lime rock of the subcarboniferous and upper Devonian periods.

The Petroleoferous rocks of the region under consideration seem to be the chloritic sand stones, which are mostly saturated with oil, and the brown shales very friable and whose layers are so loosely compacted as to admit the oil to be forced through the fissures. It is where these rocks have been exposed by the attrition of the sea, that the oil trickles out at low tide. This outcrop of Petroleum is first observed at "Mal Paso Chico" and it can be traced, at intervals, to Cape Blanco a distance of more than 30 leagues. A gooa example of oil dropping out in sand rock is seen at a distance of two and a half leagues from the sea, up the "Valle de Tuciyal," at a poit marked upon the accompanying map. A subsequent reference will be made to the richness of the vein here indicated. The extent of the Petroleoferous rocks comprehend an area of twenty leagues by seven, from "Mal Paso" to "Caña Dulce," making 140 square leagues, or 806,400 acres. This does not include the region near Cape Blanco which has not yet been examined.

## History of the Discovery of Petroleum on the Estate of Máncora.

For many years mines of Brea, Alquitran, Capé or Asphaltum, as the material has been variously denominated, have been known to exist, and have been worked at various points along the West Coast of South America in Chili, Peru and Ecuador. This Brea exists at Punta Santa Helena, in Ecuador, and at Caña Dulce, in the estate of Máncora, in two forms: on the surface it is the "Brea Piedra," Asphaltum—hard of a lustrous surface where broken and presenting a dark brown colour and conchoidal fracture. Below the surface at various depths, there exudes from sides of pits excavated for the purpose of collecting it, a black mineral tar, of a strong pitchy odour, irridéscent, readily igniting and burning with a red smoky flame entirely consumed. This is the "Brea Liquida," or liquid asphaltum, and is melted with the "Brea Piedra" to make a pitch for smearing the inner surface of *piscos* and *botijas*, for the native wine and rum. This pitch when prepared, is sold at from $25 to $30 per quintal, according to its quality. A Brea mine was opened several years ago by the proprietor of the estate of Máncora, at Caña Dulce, for the manufacture of Alquitran. As the outcrop of Petroleum, believed then to be the same with "Brea Liquida," had been observed on the shore from Cape Blanco to Mal Paso, the proprietor, in company with several other gentlemen, determined to open a "Brea Mine" at Zorritos, where the outcrop seemed to indicate a large quantity of the material. A Scotchman, by the name of Farrier took charge of this business, and commenced at Zorrites where he opened some eight pits, varying from 20 to 30 feet in length, seven to fifteen in width and sunk to a depth of from 17 to 30 feet. In most of these a surface Petroleum was obtained of a gravity varying from $910°$ to $925°$ and in quantities varying from one to five buckets per day, from each pozo or pit.

It was found, however, that this Petroleum would not answer the purpose for which it was sought. As there was no Asphaltum or "Brea Piedra" in the vicinity it would be necessary to evaporate the Petroleum until reduced to a thick tar; but the result proved to be entirely. incommensurate with the labour and expense, as but a mere residue of tar was left, nearly the whole volume of Petroleum disappearing by evaporation. As there seemed to be no other use for the Petroleum, after considerable expense, with no remuneration, the enterprise was finally abandoned. In 1863 Mr. Alexander Rudens, of Paita, for many years the commercial agent of the estate of Máncora, and interested in the previous experiment, petitioned the Peruvian Government to send Mr. A. E. Prentice, Government Civil Engineer, to examine the Petroleum Mines of Máncora.

This time the object was to extract Petroleum and the proprietor with Mr. Rudens and several other gentlemen, proposed to form a Company for that purpose. In accordance with Mr. Rudens' petition, Mr. Prentice was appointed on the 23rd September, and he procecdeded to Máncora where he collected "peons" and materials for work, and commenced operations on October 30th.

Farrier, after having abandoned the idea of making Alquitran, had collected and shipped to the Lima Gas Company about one hundred barrels of oil to determine whether gas could be made from it more economically than from coal. Mr. Prentice was also an engineer of the gas company and it was hoped that the experiment would prove successful, and thus all the gas companies on the coast become purchasers of Petroleum. Mr. Prentice proceeded to open anew the "pozos" made by Farrier and also on the 2nd of November, commenced boring a well from the bottom of "pozo" No. 4, which at that time contained 36 gallons of Petroleum of 900° gravity. The boring was continued with very imperfect tools and various success until about the middle of November when the well had reached to 79 feet 6 inches below the natural surface, and about 60 feet below the surface of the sea. The oil taken from this well stood at 842°—thermometer at 84° While engaged in this enterprise Mr. Prentice examined this section of the estate as thoroughly as the circumstances permitted, and gave the following General Opinon on its Petroleoferous character:

"The superficial deposit from which the oil has "hitherto been taken appears to be only the overflow.

"ings of large subterranean deposits, which not
"being able to come directly to the surface on ac-
"count of the impermeable nature of the immediate-
"ly adjacent strata, only at present find an outlet by
"taking a circuitous or zigzag course or by following
"the thinning out of these impermeable strata until
"meeting with the upper and more permeable beds
"of sand and loam. At Zorritos where the wells were
"dug, these permeable beds are still covered by about
"fifteen feet of impermeable variegated clays, which
"may account for the more liquid state of the oil found
"there, as compared with that found at the "Brea,"
"(Caña Dulce) about 20 leagues distant south of
"Zorritos and about 12 (?) leagues distant from the
"coast where the oil bearing strata coming to the
"surface, the more liquid or volatile parts, which is
"there collected into shallow trenches and boiled
"down to be sold as pitch. The beds of sandy loam
"from which the oil is taken being only slightly
"permeable, that is only allowing the oil to filter or
"travel horizontally through it (?) at a very low rate
"only a very small continuous supply can be expected
"from it. After the small local accumulation around
"each well has been drained, still it is not likely to
"cease altogether for a long time. If these, instead
" of having been made on the shore where the oil bear
"ing strata crops out between high and low water
"mark, (which was the circumstance which first drew
"attention to the spot) had been made a little way in-
"land the probability is that a better supply would
"have been found. The wells or pits dug at Zorritos
"were made uselessly large, 24 to 36 feet, and from
"6 to 13 feet wide, so that the cost of the oil taken
"therefrom can form no criterion as to what its ex-
"traction would cost if the pits were made in a more
"economical and systematic manner. I have no doubt

"but that a very considerable supply might be ob-
"tained, although only counting upon the superficial
"strata at from one-fifth to one-fourth the former
"outlay. I also think that there is a considerable
"probability of finding a large continuous supply by
"persevering with the boring, even although the
"the first few trials should not prove successful,

Two and a half leagues from the coast up the Tuciyal, where the oil bearing sand rock crops out as before described, a pozo was opened by Farrier seven feet square on the surface, and on reaching a depth of 28 feet the Petroleum bubbled up like a boiling spring, the oil rising to 10 feet 6 inches from the bottom in a few hours, at which depth it stood when examined by Mr. Prentice, November 1st 1863. This is probably the best surface show of Petroleum ever discovered. Mr. Prentice's report not proving so favorable as was desired for obtaining large results without involving considerable capital the organization of the Company was never perfected, and nothing more was done with the Máncora Petroleum Lands until September of the following year (1864) when a more careful survey of the land and a rough map of the coast at Zorritos and of Caña Dulce was made by your Superintendent. During the previous examinations and experiments Mr. Rudens had sent specimens of the crude Petroleum both to England and the United States for analysis.

These analyses had proved very favourably giving a small percentage of Benzine and but little residuum while the illuminating oil amounted to about 70 per cent for gases, Benzine and residuum. The analysis and specimens of the refined oil were obtained and new specimens of the oil, collected and sealed up at the wells were forwarded to New York for a fresh analysis, which were made and fully confirmed the excellent character given to this Petroleum by the former analyses. Minutes of a contract were agreed to by Don Diego Lama the proprietor of the estate, of a sufficiently favourably character to warrant the organization of a company.

The more important considerations which led to the organization of the "Compañía Peruana de Petroleo" were doubtless the following:

First.—Petroleum has been found in a considerable quantity and the extent and geological character of the territory seemed to indicate the existence of large quantities and reservoirs in store, waiting to be developed and utilised by enterprise and capital.

There is one geological characteristic which has been reserved for present consideration and that is the dip of the rock. Numerous volcanic disturbances have here and there bent and distorted the strata in almost every direction; but for leagues along the coast and in every "quebrada" inland the strata are seen to dip uniformly when otherwise undisturbed to the southeast at an angle varying in different localities from 15° to 40°.

As the strata are tilted up towards the coast and under the sea, their edges exposed by the removal of the debris, furnish the remarkable] outcrop of Petroleum beforementioned and, what is of more importance, prove that it must have come from a great depth pressed up between the strata to the surface by its accompanying gases. Thus it was observed in all the pits that were dug. that the Petroleum flowed in from the lower or southeast side and by taking the outcrop upon the shore and tracing it according to the angle of the dip, back along the strata to a well, it was found that the depth at which the vein would be struck could be calculated with great precision.

Second.—Peru, where the territory is located, furnishes a ready home market for a large quantity of refined Petroleum.

In March 1861, the first gallon of refined Petroleum introduced into the country for public use was exposed for sale in the city of Callao; and now the annual consumption reaches to a million gallons; and if Chili and Ecuador be included it will increase the amount to a million and a half. All this is brought from the United States and has to pay a revenue duty amounting in Peru to twenty-five cents per gallon. The amount of the article consumed is rapidly increasing and this would no doubt continue for several years could the price be reduced so as to extend its use to the poorer classes. Australia is within forty-five days by sail, with constant south east trades for the return passage as well. The Australasian British Colonies consume more than three million gallons of refined Petroleum annually almost all of which is imported from the United States.

Third—To say nothing of California and Oregon, which are convenient markets should they not succeed in supplying their own Petroleum, it is a principal consideration that this territory lies stretched for ninety miles along a coast indented with good harbors where the Petroleum can be produced and shipped at the least possible expense. The wells already sunk are within 200 feet of high water mark thus avoiding all expense of land carriage, so that the oil produced by these wells or any others similarly located on the coast can he furnished to the London or Liverpool market for less than half the expense of the Pennsylvania oil. This fact opens the European market to the products of this territory. The Company has been assured that as soon as a constant supply of oil can be produced to warrant their undertaking, a line of iron ships with large iron tanks to carry the oil in bulk, will be built for this trade. Such were the considerations on which the company was organised. Two points remained to be settled by actual experimeut upon the ground to warrant the expenditure of the large capital necessary for the development of the territory, the cost of producing the oil and the quantity that could be obtained. In order to settle these questions a prospecting party was sent out from New York in August 1865 with a single engine and two sets of drilling tools. The party arrived in September. Everything had to be made and prepared de novo. The party lived under an old shed on the beach until the men's quarters and storehouse could be built; and on the 30th October drilling commenced. Various strata of clay shales, sand stones, and conglomerates varying from three inches to two feet in thickness were passed through down to 56 feet 6 inches where a small vein of salt water was struck and at 70 feet in loose conglomerate the first considerable vein of oil and gas was met with. In the alternate layers of shale sand and conglomerate, small veins of oil continued to be encountered every few feet. These were not crevices but like the small veins found in sinking a well for water. From 107 to 114 feet the gas and oil improved rapidly and it was thought best to tube and test the well. Tubing was accordingly put down fo 111 feet. After exhausting the water, the well pumped at the rate of a little more than two barrels per hour for forty hours; but as no tankage had been prepared in advance, it was thought best to sink the well deeper with the hope of finding a large crevice and a flowing well. It would be dfficult to say how many good wells have been spoiled by not letting well enough alone. At 190 feet a mud vein was struck which filled the well to within 70 feet of the surface and so No. 1 was lost. Well No. 2 was commenced about the 20th December and passed down through a somewhat firmer rock although of the same general character, excepting that but little conglomerate was found. The same oil bearing strata were encountered as in No. 1.

No oil was found between 110 and 160 feet, but from 160 to 185 feet there was a good show. At 185 feet the rock growing soft drilling was suspended and the well tubed to 177 feet. No. 2 has pumped and flowed alternately ever since averaging about 25 barrels per day. It has not diminished in quantity and it is only necessary to pump it 2 hours daily.

Well No. 2. was put down at a distance of about 100 feet from No, 1. and No. 3. was commenced at 113 feet from No. 2. along the coast line. In this well the upper veins were more abundant, and at 56 feet four barrels a day could be dipped off regularly with the sand pump. From 56 feet to 132 feet no oil was found; but from 132 feet to 422 a good show of oil and gas was almost constanly met.

From 422 to 463 feet hard lime rock was encountered. gradually growing softer and more argillaceous, and finally changing to a loose bituminous shale, with an excellent show of gas and oil. This continued, gradually increasing, to 526 feet where an oil crevice of a foot in depth was struck. The well was put down to 531 feet and then tubed and tested.

It proved a great misfortune that the only spare pump barrel was seriously damaged by being bulged in the middle; the packing that fitted the middle of the barrel was too tight for the ends, and would strip off; and when properly packed for the ends it was too loose for the middle. There were no facilities at hand to remedy this defect satisfactorily; but after drawing tubing five times, the pump worked indifferently for several hours with the following result:— Four hours were spent in exhausting mud and water when the well commenced pumping and flowing oil at the rate of 480 barrels per day; this continued until the packing came off and stopped everything. In drawing tubing at 250 feet from the bottom, the bore-chip, which caused the difficulty, shook loose, and the well commenced flowing in a stream the full size of a two inch pipe, which continued, although the seed bag had been drawn, until the last length of tubing was withdrawn from the well.

After the tubing had been removed, it was determined to drill the well to five hundred and fifty feet, to allow ample space for the chips and debris of boring to settle below the crevice. In attempting to accomplish this object, the tools became fast in well, and every effort to remove them has thus far proved unsuccessful.

Well No. 4. has been commenced a short distance from No.3. and was down to 109 feet Oct: 27th, with a show of Oil and Gas, corresponding thus far with No. 3. The accident to No. 3. disappointed the expectations but did not discourage the efforts of the Company.

The results of the last year's prospecting have convinced the Company of the great extent and permanency of the Petroleum deposits and that wells can be bored and pumped as cheaply on the estate of Máncora as in Venango Co. Pennsylvania, while extensive markets are waiting to purchase the oil as soon as produced.

With these views the Company has secured the services of Mr. Geo. E. Corey (a man of extensive experience in putting down wells in Pennsylvania; and under whose supervision as chief borer the prospecting of the last year was conducted), and has supplied a corps of experienced workmen.

Three additional engines have been purchased, a small steamer built for supplying water and provisions to the works, and a bulk boat for loading and discharging cargo; also 8,000 barrels of iron tankage and all tubing and machinery necessary to secure the complete success of the enterprise have been provided.

The Máncora oil territory has been carefully examined by men familiar with the Pennsylvania oil Lands. Mr. Corey has sunk sucessfully more than sixty wells in the best oil lands in the United States, and your Superintendent made a careful examination of the oil Creek territory before securing the contract for the Máncora estate; and it is his deliberate opinion, as well as that of Mr Corey and of every other practical man who is thoroughly acquainted with both localities, that were both to-day virgin territories untouched by the drill, the Máncora estate has a better show, a greater extent, a more favourable location to supply the great marts of the world, and a better prospect for a large and constant yield, of oil, than the entire Pennsylvania oil region. It only remains to demostrate the correctness of these views by the actual development of the "Peruvian Petroleum Co's Oil Lands",

It is confidently hoped that before the time for another annual report shall arrive, your Superintendent will be able to lay before you results that shall satisfy every reasonable expectation.

I have the honour to be, very respectfully, your obedt. servant,—E. P. LARKIN.—General Superintendent.

CONDENSED REPORT FURNISHED BY W. WARREN, ESQ.
M.I.C.E., F.R.G.S.
1887.

The estate extends a distance af over 30 miles along the sea coast of the Pacific Ocean, being situated in the most northerly province of Peru, 18 to 53 miles north of Payta, and at Talara, a point about central of the oil belt of this estate; there is an excellent deep-water harbour, with good anchorages, well sheltered, and otherwise offering extremely favourable conditions for the provision of every facility and requirement for dealing with the largest possible shipments of petroleum, both crude and refined.

Futher, the estate occupies a position in relation to the whole west coast of South America, as indeed of North America, most favourable in every respect for supplying not only those western countries, but also countries farther west, such as New Zealand, Australia, Japan, China, etc., the saving in distances from the east coast of the U. S. A. to these latter countries being as much as 5,500 to 8,000 miles. For the consumption also of the west coast of America the saving in distance is very important, the present supplies having to come by way of Cape Horn or the Straits of Magellan and not by the trans-continental railways.

Petroleum oil abounds throughout the entire coast length of the estate and is found in great abundance over large areas from the coast to 10 or 15 miles inland, from which there would appear to be ample reserves to last for generations. Some of the wells are placed within a few hundred feet of high-water mark.

Like the Russian oil-fields, the La Brea oil territory has been known and worked for hundreds of years, chiefly, however, for the manufacture of pitch, by evaporating off the lighter oil products and petroleum or kerosenes, the resulting pitch being used, from ancient times to the present, for glazing the inside of earthen jars used by the native Peruvian and Chilian wine producers.

Some 16 wells have been drilled at different and distant points on the estate, six quite recently; nearly all are productive at present, although one—now producing good oil—was drilled some 20 years ago, and another, drilled some 14 years ago, is still productive and has yielded an average of over 30 barrels daily for that period. This well gave 70 barrels daily at first, though now reduced by neglect of former owners to 10 barrels.

Other wells have given as much as 400 barrels each daily, and the six wells drilled during the present, year are good for an average of say 200 barrels each daily, or an aggregate for the six wells of 1200 to 1500 barrels, worth say £ 240 to £ 300, equal to £ 72,000 to £ 90,000 yearly.

The operation of well-drilling is very easy, one good productive well 210ft. deep having been drilled in less than five days, while the average for the six wells has been 345ft. and ten days respectively. The average depth in America and Russia is nearer 1,000ft., and the cost per foot much greater.

Everything points to the existence of enormous supplies of petroleum obtainable from the different oil belts upon the estate, one of which, at Negritos, some two square miles in area, or 1,280 acres, having been explored by drilling 12 wells, *none of which failed to produce oil.*

The two most distant wells at Negritos are nearly three miles apart, the intervening land having been more generally worked, as being equally productive. These wells are 12 miles west of the equally productive wells drilled upon the oil belt inland at La Brea, between which and the wells referred to as at Negritos, there are other extensive but as yet unexplored oil belts. Should this area of two square miles at Negritos be only half as productive as an equal area of 1,329 acres at Bakhu is stated to be (14.375,000 barrels being credited to it for 1886) it would give 7.000,000 barrels of crude oil, value say £1.000,000; while, even if the output only equalled the best results obtained from a similar area of America oil territory (Venango) which gave 1.000,800 barrels, value say £ 300,000, the return would still be very satisfactory. Inasmuch, however, as the wells hitherto drilled on the La Brea estate at Negritos show results far supassing the average yield obtained in America, if not in Russia, averaging about ten times more than an equal number of wells of the American oil fields, the output for the two square miles at Negritos may be estimated at 15:000,000 barrels, or a value of say £3.000,000.

The oil territory of America is returned as 1,339 square miles of which however only some 39½ square miles, on which have been drilled 22,524 wells, have produced the bulk of the output; and similarly at Bakhu 1229 acres, or less than two square miles of the estimated area of 1,200 square miles, containing 400 wells, have produced the bulk of the Russian oil.

The average production of oil per square mile worked in America down to 1885 is returned as 740,000 barrels, against an incomparabily greater production in Russia; and judging from the results already shown by wells on the La Brea estate, the output there will greatly exceed the square mile production of America.

In America the average oil production per well has averaged about 12,000 barrels as compared with 76,000 barrels per well in Russia, and estimating the production from the two square miles of the oil belt on the immediate foreshore of Negritos at 50,000 barrels per well, there would be, from 300 wells, 15.000,000 barrels of an estimated value of 3.000,000 without considering the certainly ten times larger oil areas adjacent thereto.

Fity wells may be drilled in a year, which would give a large and prompt return, both from the sale of crude unrefined and refined oils. The former, which would yield a larger profit than the refined oil, will probably be sought after and meet with a ready market upon the Pacific coast of North and South America, for gas work, copper smelting, nitrate evaporation, water condensing, local oil refineries and other works, as it has calorific value of from two to three times more than coal, and also produces more and better gas, even in common coal retorts, than coal, which has to be imported from Europe.

The character of the oil is brilliant in every respect, the produce of the last well drilled having yielded 91.5 per cent. of oil, which might almost all be classified as an illuminating oil, no distillate heavier than 8.114 specific grauity, or 42° Beaumé, having been obtained.

The "Flash" point of the kerosene averaged about 100° F.—as obtained by laboratory distillation, and without any treatment—being somewhat low; but on the large comercial scale and usual treatment, the flash point can, I believe, he made to conform to any requirement up to 130° F.

As will be seen from my full report giving all details as to distillations, the lighter oils began to come off at 115° to 160° F., and with a temperature of 700° F., about 55 per cent. of all the distillates had come off of a bright water white character which may be at once burnt in an ordinary lamp giving a brilliant light, without chemical or other preparation or treatment of any kind.

W. Warren.

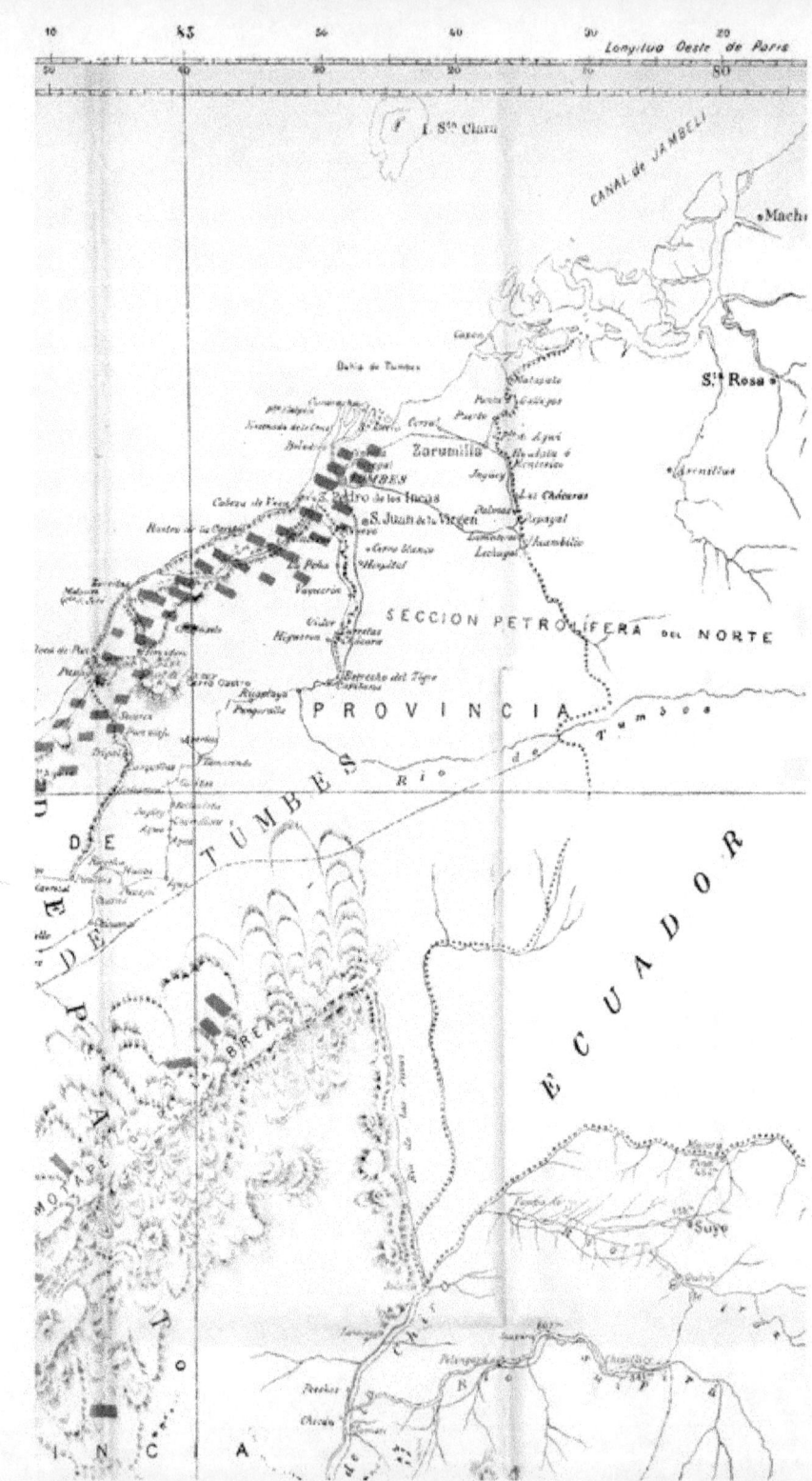

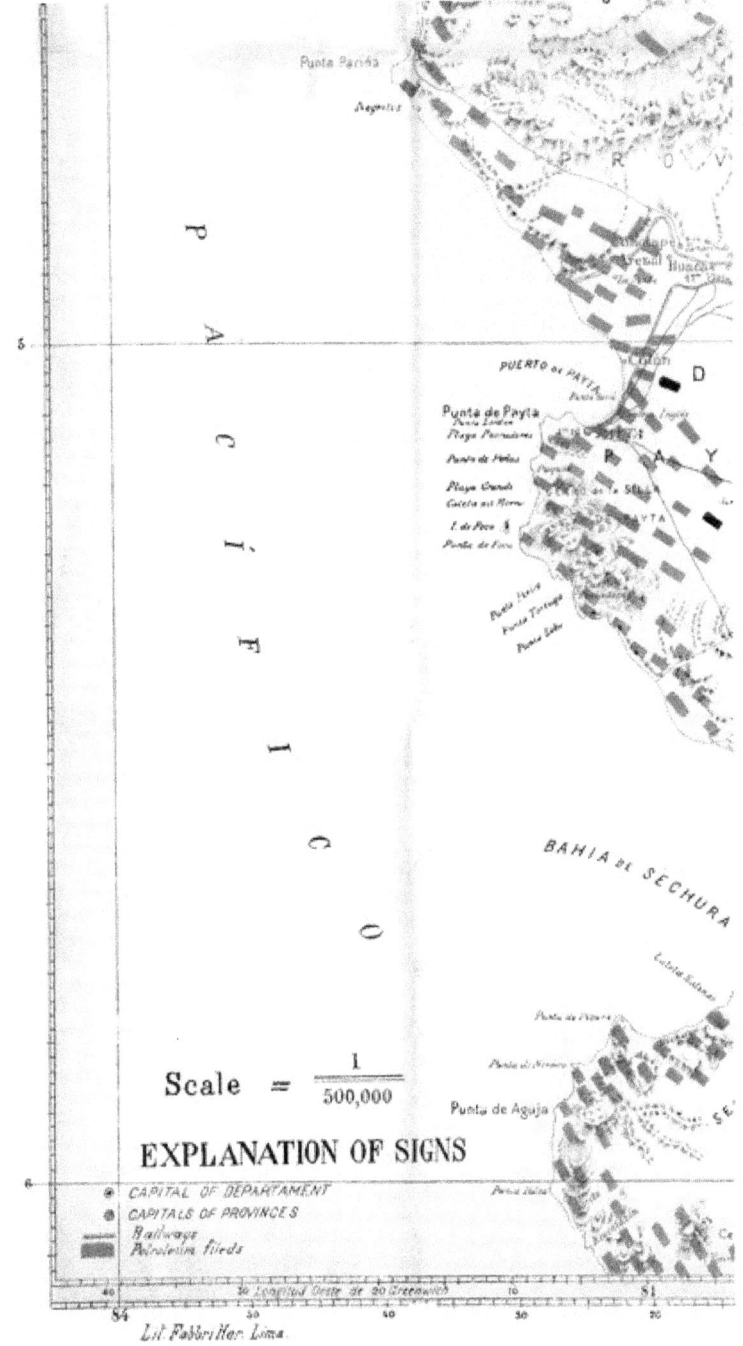

# OTHER WORKS BY THE SAME AUTHOR.

The Tobacco, Salt and Alcohol Monopolies.
The Erection of the Monument in Honor of Rear-Admiral Grau, in Piura, 1886.
Defense of the Sears Irrigation Project, 1887.
The Tobacco Monopoly in France, 1886.
Memoranda for a Report on the Department of Piura, 1890.
Reply to the Letters from London on Peruvian Mines, 1891.
Petroleum in Peru (Spanish original), 1891.

## FORTHCOMING.

Second Edition of the Report on the Department of Piura.
Technical Data on Peruvian Petroleum.
Peruvian Quicksilver Mines.
Comparison between the Returns of Mines in Peru, and the Mines in other Countries in America.

www.ingramcontent.com/pod-product-compliance
Lightning Source LLC
Chambersburg PA
CBHW031452160426
43195CB00010BB/954